Springer Studies in the History of Economic Thought

This series offers an outlet for research in the history of economic thought. It features scholarly studies on important theoretical developments and great economic thinkers that have contributed to the evolution of the economic discipline.

Springer Studies in the History of Economic Thought (SSHET) welcomes proposals for research monographs, edited volumes, textbooks and handbooks from a variety of disciplines that seek to study the history of economic thinking and help to arrive at a better understanding of modern economics. Relevant topics include, but are not limited to, various schools of thought, important pioneers and thinkers, ancient and medieval economic thought, mercantilism, cameralism and physiocracy, classical and neoclassical economics, historical, institutional and evolutionary economics, socialism and Marxism, Keynesian, Sraffian and Austrian economics, econometrics and mathematical studies as well as economic methodology and the link between economic history and history of economic thought.

All titles in this series are peer-reviewed.

For further information on the series and to submit a proposal for consideration, please contact Johannes Glaeser (Executive Editor Economics) Johannes.glaeser@springer.com.

Carlo Zappia

Uncertainty in Economics

A History

Springer

Carlo Zappia
Department of Economics and Statistics
University of Siena
Siena, Italy

ISSN 2662-6098 ISSN 2662-6101 (electronic)
Springer Studies in the History of Economic Thought
ISBN 978-3-031-91220-7 ISBN 978-3-031-91221-4 (eBook)
https://doi.org/10.1007/978-3-031-91221-4

Translation from the Italian language edition: "L'incertezza in economia: Una storia delle teorie da Keynes ai giorni nostri" by Carlo Zappia, © Author 2024. Published by Carocci editore. All Rights Reserved.

Previously published by Carocci editore

This Springer imprint is published by the registered company Springer Nature Switzerland AG
The registered company address is: Gewerbestrasse 11, 6330 Cham, Switzerland

Acknowledgments

The present volume constitutes a revised and enlarged version of my "L'incertezza in economia. Una storia delle teorie da Keynes ai giorni nostri," published in Italian in 2024. It concerns topics that have constituted the focus of my research for many years. My first thanks, therefore, are due to my colleagues who were co-authors of some of the research that provided the inspiration for this study: Marcello Basili, Alberto Feduzi and Jochen Runde. I must also thank some colleagues who have frequently read my previous work and who were gracious enough on this occasion to read and comment on the entire manuscript in its earlier Italian version: Jack Birner, Massimo Di Matteo, Nicola Giocoli, Mario Pomini, Enrico Petracca and Roberto Romani. Other colleagues provided equally valuable feedback on specific chapters: Valentina Erasmo, Luca Fiorito and Ivan Moscati. The comments made by Anna Carabelli, Giacomo de Gioia, Alessandro Innocenti, Cosimo Perrotta and Salvatore Rizzello after the publication of the Italian version proved to be of great help in the process of drafting this new version, which also benefited in its final revision from the kind comments of George Bent, Mauro Caminati, John Davis, Andrew Freeman and Itzhak Gilboa.

Writing a book absorbs much of the author's energy and attention for a period of time, so the support of those close to you is particularly important. The support that I have received from my wife Patrizia Lattarulo and our children Antonella, Daniele and Francesco has been paramount. But equally important over the years has been the kindness shown to me, first in a face-to-face meeting in Siena in 2005, and then in a series of subsequent email exchanges, by Daniel Ellsberg, a scholar whose role is crucial to the story told here, but also a true hero of post-World War II American history, who recently passed away. This book is dedicated to his memory.

Contents

Chapter 1
Introduction

> By 'uncertain' knowledge, let me explain, I do not mean merely to distinguish what is known for certain from what is only probable. The game of roulette is not subject, in this sense, to uncertainty; nor is the prospect of a Victory bond being drawn. Or, again, the expectation of life is only slightly uncertain. Even the weather is only moderately uncertain. The sense in which I am using the term is that in which the prospect of a European war is uncertain, or the price of copper and the rate of interest twenty years hence, or the obsolescence of a new invention, or the position of private wealth-owners in the social system in 1970. About these matters there is no scientific basis on which to form any calculable probability whatever. We simply do not know.

John Maynard Keynes, the most influential economist of the twentieth century, identified uncertainty and the difficulty of representing it as one of the key issues in his seminal work, *The General Theory of Employment, Interest and Money*.

The sentence quoted in the epigraph, taken from a 1937 article intended to summarize the meaning of that work, is one of his most frequently quoted remarks. In defending the *General Theory* against its critics, Keynes (1937, pp. 213–214) pointed out that in an economic system employment depends on the level of national income, which in turn is primarily determined by aggregate investment. However, investment decisions are heavily influenced by expectations of future returns, which are inherently uncertain because they can be affected by the most unpredictable events. Consequently, a correct theoretical analysis of economic behavior depends on the capacity to explicitly account for uncertainty, an issue that Keynes ([1936] 1973, p. 1) perceived as having been disregarded by 'classical' theory—as he considered all economic theory before him.

The issue is well summarized in his correspondence with one of his collaborators. In a letter to Hugh Townshend of April 1937, Keynes ([1937–1938] 1979, p. 258) wrote:

"I think there is a great deal to be said for your view that the classical theory is not valid on any assumptions, and that you have put your finger on the spot in saying

C. Zappia, *Uncertainty in Economics*, Springer Studies in the History of Economic Thought, https://doi.org/10.1007/978-3-031-91221-4_1

that they [the classical economists] are trying to describe a world in which risk exists without uncertainty."

It is therefore fair to say, as have his followers, especially those known as Post Keynesians, that for Keynes the revolutionary character of his theory lies more in the explicit consideration of uncertainty as distinct from risk than in other aspects.

As we shall see, the distinction between risk and uncertainty in economics is consolidated according to a scheme introduced by Frank Knight in his 1921 book *Risk, Uncertainty and Profit*. According to Knight, there are events for which it is possible to identify a group of analogous cases and thus to derive a probability of the occurrence of the single event, as in the case of the frequency of a given outcome generated by a random mechanism—for example, red in the game of roulette—or of an event similar to a series of others grouped together by homogeneity—for example, a fire at a house located in a specific spatial and temporal context. For other events, however, especially those characterized by being essentially unique—such as the future returns of a new business activity—this is not possible. In such cases, the best one can do is to make generic estimates, but it is not possible to reason in terms of the true probabilities of the events that might affect the outcome of a given activity. Knight ([1921] 1964, p. 20) therefore distinguished between "measurable risk" and "unmeasurable uncertainty."

The similarities between the distinction proposed by Knight and that suggested by the various examples in Keynes's sentence are obvious. In order to arrive at the identification of the extreme uncertainty that characterizes some events of which "we simply do not know," Keynes starts with cases that are merely risky and then outlines a series of further cases for each of which it seems increasingly difficult to assign a "calculable" probability, such that they are therefore more and more uncertain. It is thus possible to identify a unique strand of research on these issues, although economists traditionally speak of Knightian uncertainty.

However, it is noteworthy that Keynes was also the author of a seminal study on probability, which he completed during his youth and prior to the publication of the *General Theory*. In his 1921 volume *A Treatise on Probability* he argued that it is necessary to study in probabilistic terms even those situations where there is no possibility of using a probability derived from the frequency with which phenomena occur. Keynes's objective was to examine the probability inherent in the ordinary reasoning of individuals—the reasoning by which opinions or beliefs are expressed—a subject, he believed, that probability scholars of his time had neglected. His conception of probability is intricate to articulate, as it cannot always be quantified with precise numbers, yet it can be rationalized. As we shall see in detail, it is Keynes's ideas on probability that we need to start from in order to understand the notion of uncertainty that he later used in the *General Theory*. It is on the foundation of Keynes's theory of probability that a distinct Keynesian analysis of uncertainty emerges, which is only partially equivalent to the Knightian one.

Keynes is, of course, much better known for his macroeconomic analysis and his innovative ideas on economic policy, which inspired the interventions of numerous governments. These governments used Keynes's ideas to bring about the recovery of Western economies after the Great Depression of the 1930s and to sustain the long period of growth after World War II. Moreover, after decades of neglect of his economic policy prescriptions, largely due to their inadequacy in tackling the stagflation that characterized the early 1970s, the recent Great Recession triggered by the financial crisis of 2007–2008 has prompted a renewed interest in those Keynesian economic policies that effectively addressed the crisis of the 1930s.

However, the issue of uncertainty is also crucial in this respect. In the *General Theory*, Keynes ([1936] 1973, p. 148) insisted that the "state of long-term expectation" determines the ability of entrepreneurs to maintain an adequate level of investment for the full utilization of resources. Investment is the driving component of effective demand; therefore, if private investment fails due to the uncertainty that characterizes a given economic situation, only public investment can revitalize the system. In an economic system that is not guided by economic policy, it can be difficult to have expectations that are compatible with a level of employment close to full employment. The necessity for economic policy measures—especially fiscal rather than monetary—is thus closely related to the remark that, when uncertainty is pervasive in the system, it is misleading to analyze it as if it were a risk, which, for Keynes, was the weakness of 'classical' economics.

However, an examination of the recent history of economics—a history in which economics has become increasingly formalized and redefined as an exact science rather than a social science—reveals that the distinction between risk and uncertainty has been substantially trivialized, if not outright denied. In fact, while Keynes was reaffirming the importance of the distinction, a new generation of mathematical economists began to take their cue from John von Neumann and Oskar Morgenstern's *Theory of Games and Economic Behavior* to delineate a treatment of uncertainty according to the canons that had been defined for risk within game theory. In the second edition of their book, von Neumann and Morgenstern ([1947] 2004) provided a mathematical justification of the criterion of maximizing expected utility as a criterion for rational choice, proposing to weigh the utilities associated with the possible consequences of the chosen action against the probabilities with which those consequences are expected to materialize. In the modern interpretation of games, this applies to the behavior of the other players, but by analogy it can be thought of as applying in all cases where there are external events that condition the payoffs of the decision-makers, such as in board games, and generally in all situations where the probabilities of payoff-relevant events are assumed to be known.

We shall see that the expected utility criterion had already been considered, albeit unsystematically, by nineteenth-century economists such as William Stanley Jevons and Alfred Marshall, among others. Philosophers of probability and mathematical statisticians had also treated the question in these terms earlier, between the seventeenth and eighteenth centuries. This was done implicitly by Blaise Pascal, but more clearly by Daniel Bernoulli. However, it was only after the result of von Neumann and Morgenstern—a mathematical theorem that established a precise link between rationality, defined by a few axioms, and the employment of an 'expected' utility function—that the criterion of maximizing expected utility became universally accepted for rational choice theory. Indeed, it became so accepted that it was generalized beyond its original context. Although designed for risk—situations where probabilities are always given, as if dependent on a random mechanism—expected utility maximization was soon applied to Knightian uncertainty on the basis of a specific subjective view of probability.

The hypothesis that subjective probability, that is, the probability proper to individuals rather than to the external world, can always be measured numerically, emerged from a critical reading of Keynes's work by Frank Ramsey, who was a young associate of Keynes in the years when the *Treatise* was published. According to Ramsey, recourse to ordinary reasoning to determine the probability of an uncertain event must necessarily imply that one can infer from actions the probabilities that justified them, and a common method of deriving this information is to ask individuals to bet on events to which it seems impossible to assign any meaningful probability, as in the case of horse racing. In this way, Ramsey argued, it is possible to infer the decision-maker's probabilities even when they appear to be based on undefined degrees of belief.

In an even more elaborate way, both philosophically and statistically, this theme was independently developed by Bruno de Finetti in his work of the early 1930s. De Finetti identified the conditions under which subjective beliefs are numerical probabilities, with all the typical properties usually ascribed to probability functions, starting with the possibility of summing the probabilities of disjointed events to identify the probability of their union, according to the so-called property of additivity. Although only indirectly known in the Anglo-Saxon world until the early 1950s, de Finetti's ideas—which largely coincided with those of Ramsey—formed the nucleus around which a theory of decisions with subjective probabilities was constructed that had the same explanatory power as that with objective probabilities. The elaboration of de Finetti's approach proposed by Leonard Savage in *The Foundations of Statistics* ([1954] 1972) eventually showed the mathematical conditions under which what von Neumann and Morgenstern had proved for situations of risk also holds for situations of uncertainty. Since then, almost all economic studies have treated uncertainty as risk.

The history of uncertainty presented in this volume therefore necessarily moves between two extremes. On the one hand, there is the emphasis placed by the most important economist of the last century on the disruptive effects of uncertainty on the outcomes—even the aggregate outcomes—of individual choices. Such an emphasis stems from considerations of the problematic nature of its representation in a probabilistic context. With a typical political economist's approach, Keynes proposed the use of models to represent uncertainty, but he set limits on their use, making it clear that aspects that cannot be mathematically formalized should not be excluded from economic inquiry. On the other hand, a well-defined set of techniques is entrusted with the task of outlining a modern economic science capable of prescribing precisely how agents should behave in any situation, even the most uncertain. For the analysis of choices conditioned by events that are largely neither repeatable nor comparable to others—choices to be made under uncertainty rather than risk—a decision-making framework typical of strict economic rationality is used, based on consistency constraints between the choices made by the economic agent. From the 1950s onwards, what was already common in the theory of choices under conditions of certainty also became common for choices under both risk and uncertainty.

The clash between these two visions is still paradigmatic in economics today. More than on many other issues, economists are divided on how to deal with uncertainty. The majority of economists understand economics as a formalized science capable of providing unambiguous answers in any context, sometimes even with a predictive character, and argue that rational choice theory is all-encompassing and only needs to be adapted to different applications, including contexts where knowledge is considerably incomplete. This dominant view is countered by a group of political economists—less quantitatively relevant but no less important—who deny that developments in rational choice theory allow uncertainty to be reduced to risk. Based on the teachings of Knight and Keynes, what we might call the economists of irreducibility (of uncertainty to risk) advocate a rationality understood in a broad sense, to distinguish it from the narrow rationality of the dominant theory. Following Keynes, these authors do not object to the construction of models that specify the limits of the analysis and the relationships between the variables used. However, they reject simplification, especially when the object of analysis is complex, as is the case in all situations where uncertainty is relevant.

A history of economic ideas and theories that foregrounds the theme of uncertainty can therefore be useful in characterizing the origins and deeper meaning of a dichotomy between modes of analysis that is still relevant in economics today. In a world of sudden financial crises and looming climate change, of pandemics that were thought to be containable and wars that were hoped to be only local, uncertainty is once again a major topic of discussion. This is so not just among 'experts,' but in economic, political and social discourse more generally. The manner in which uncertainty is represented and the tools employed to address it have a history that, at least from an economic point of view, shapes the answers we seek today in the face of phenomena of such importance. The aim of this volume is to present this history in a concise yet comprehensive manner.

The book is structured as follows. Chapter 2 introduces some basic concepts pertaining to the interplay between probability and risk, as outlined in the first—not always systematic—studies of the subject. The insights of Pascal and Bernoulli, who were among the earliest to employ mathematical analyses of games of chance, are used in conjunction with the analyses of Jevons and Marshall to define the field. Chapter 3 delineates the distinction between risk and uncertainty, following the approaches of Knight and Keynes. This is primarily done with the aim to highlight the commonalities in their approaches, while also acknowledging the differences between Knightian and Keynesian uncertainty that emerged in Keynes's *General Theory*. Chapter 4 reconstructs the impact of von Neumann and Morgenstern's version of expected utility theory and illustrates its integration into the axiomatic approach developed in the 1940s. Harry Markowitz's attempt to justify portfolio choice on the basis of expected utility is also briefly discussed. Chapter 5 provides a comprehensive account of the subjectivist approach that emerged as the predominant paradigm in economics in the second half of the twentieth century. After an analysis of the contributions of Ramsey and de Finetti, the focus is on the reasoning scheme employed by Savage to incorporate decisions under uncertainty within a probabilistic framework that essentially coincides with that used for risk. Chapter 6 presents the so-called paradoxes of choice. The chapter places particular emphasis on Daniel Ellsberg's introduction of the notion of ambiguity, a concept that has played a crucial role in generating a growing unease with the dominant approach even among economists specialized in decision theory. Chapter 7 aims to demonstrate the continuing relevance of Keynes's message, both in the post Keynesian studies of 'radical' uncertainty and in those contemporary decision theories that challenge—even in terms of mathematical representation—the idea that there is no need to differentiate between risk and uncertainty. The so-called return to Keynes after the Great Recession is also highlighted.

This volume offers a distinctive interpretation of uncertainty that, while placing Keynesian uncertainty at the center of the analysis, does not intend to neglect the importance given in modern economics to a mathematical representation of uncertainty. However, it is important to emphasize that it is written in the form of a single-subject textbook, and it is intended to be accessible even to readers who are not expected to have any special knowledge other than that provided by an introductory course in microeconomics. The book is therefore suitable for use in courses on the history of economics and economic thought, also as an aid to understanding issues that are often only hinted at in generalist textbooks. Uncertainty issues are, in fact, crucial to understanding the development of economic theory in the twentieth century. For this reason, the volume may also be useful in providing a more problematic and historically motivated view of topics covered in courses typically offered in Masters Programs in Economics and Business studies, such as advanced microeconomics, decision theory, finance, and game theory and information.

References

Keynes, J. M. ([1936] 1973). *The general theory of employment, interest and money. The collected writings of John Maynard Keynes* (Vol. VII). Macmillan.

Keynes, J. M. (1937). The general theory of employment. *Quarterly Journal of Economics, 51*, 209–223.

Keynes, J. M. ([1937–38] 1979). Letters to Hugh Townshend. In Id. *The general theory and after: A supplement: The collected writings of John Maynard Keynes* (Vol. XXIX). Macmillan.

Knight, F. H. ([1921] 1964). *Risk, uncertainty and profit*. Augustus M. Kelley.

Savage, L. J. ([1954] 1972). *The foundations of statistics*. Dover Publications.

von Neumann, J., & Morgenstern O. ([1947] 2004). *Theory of games and economic behavior* (2nd ed.). Princeton University Press.

Chapter 2
The Art of Conjecturing: Probability and Utility

2.1 Introduction

Current economic theory is largely based on an established model of individual decision-making. The economic agent, often referred to as the decision-maker, formulates choices that, given a preference order over possible outcomes, are efficient in the sense that they determine outcomes that cannot be improved upon, within a framework defined by external constraints. Traditionally, at least since the period of the so-called marginalist revolution in the years around 1870—which for the sake of simplicity we will identify with William Stanley Jevons's approach, but which of course also refers to Léon Walras's theory of general economic equilibrium and Carl Menger's investigation of economic needs as subjective values—the problem of choice has been considered as a problem of maximizing the utility that the individual agent can derive from the possible outcomes. The most classic example is that of consumer choice, where different baskets of purchasable goods are evaluated in terms of their associated utility, given a budget constraint. Similarly, the entrepreneur is regarded as a profit maximizer who compares the revenues that can be obtained from the production of goods with the costs incurred in producing them, given a technology that relates inputs to outputs.

Among the early marginalists, Jevons was the one most influenced by Jeremy Bentham's utilitarian philosophy, according to which all human action is guided by the pursuit of pleasure and the avoidance of pain. Within this philosophical approach, which was widely shared by the economists of the time, utility is a sort of psychophysical entity that can be measured in terms of the intensity of a sensation; consequently, the utility function is a yardstick for comparing the pleasures and pains of economic action, as if it were possible to measure and compare its effects. Since utility is supposed to be measurable, any increase in utility can be compared with the cost, in terms of disutility, of obtaining it. If we focus on the utility associated with the consumption of a basket of goods, the increase in utility obtained by consuming an additional unit of one of the goods in the basket is a number which, when related

C. Zappia, *Uncertainty in Economics*, Springer Studies in the History of Economic Thought, https://doi.org/10.1007/978-3-031-91221-4_2

to the price to be paid to obtain that additional unit of the good, can be compared to the increase in utility obtained by consuming an additional unit of another good, again related to the price to be paid to obtain it. A given amount of money would be allocated to the good whose ratio of utility increase to price is greater. Marginal utility thus became the fundamental concept of economics at the time.

Later, however, at least since Vilfredo Pareto's (1896) analysis at the end of the nineteenth century, this specific utilitarian approach was abandoned. Economists began to admit that utility, as a measurable entity, was a concept devoid of empirical content. After Pareto, the question became rather one of identifying the canons of rationality that must guide choices, regardless in principle of their representation through the utility function. The utility function, intended as a function to maximize, remained central, but utility began to be used only as a synthetic index representing the preference order on the objects of choice. Utility was no longer used to measure their absolute importance, let alone to make comparisons between variations at the margin. Pareto inaugurated the modern view, according to which preferences are represented by individual indifference curves, and the marginal rate of substitution becomes the fundamental notion for evaluating the increase in individual welfare that results from consuming one more unit of the good. This increase is now measured in units of the other good that an individual is willing to give up, rather than in terms of marginal utility. This is the approach that still characterizes the teaching of microeconomics in today's textbooks.[1]

Of course, today's economists admit that it is an extreme simplification to assume: (a) that decision domains are precisely delimited; (b) that individuals have well-defined preferences that do not change during the decision process; and (c) that computational aspects may not be discussed despite their obvious complexity. But the reliance on a theory of rational choice has been considered crucial by those economists who, at least since Lionel Robbins's (1935) classic definition, interpret economics as a science that is concerned with identifying the means to achieve efficiently the ends taken as given.

To make more explicit this character of economics as a science that proceeds by abstraction and through the use of models, the contemporary economist usually specifies that the analysis takes place at a normative rather than a descriptive level. Economics is concerned first with analyzing what individuals should do in a simplified theoretical context—what 'norm of behavior' they should follow—and only later with what they would do in a complex real-world context. Moreover, the relevance of rational choice theory is in any case subject to 'empirical' checks, because the dominant approach assumes that individuals' actual choices can be assessed in terms of their consistency, as in revealed preference theory, or by constructing controlled decision contexts, as in experimental economics.[2]

---

[1] The distinction we refer to is evidently that between 'cardinal' utility, a notion used more or less implicitly by all marginalist economists until the end of the nineteenth century, and 'ordinal' utility, a notion associated with the use of indifference curves. On the issue of the measurability of utility we will elaborate in more detail in Chap. 4, but for an in-depth analysis see Moscati (2018).

[2] The distinction between the descriptive and the normative (also known as prescriptive) aspect became crucial at least from the 1940s onwards, the years of the birth of an 'exact' economic

The extent to which the reasoning structure just described is appropriate even when the result of the choice is not a precise outcome, but a set of possible outcomes conditioned on an external factor that does not depend on the decision-maker, is the subject matter of the extension of rational choice theory to the realm of uncertainty. Following the redefinition of the context of choice under conditions of certainty in accordance with the principles set out by Pareto and the subsequent interpretations of his approach by prominent figures such as John Hicks and Paul Samuelson—widely regarded as the leading proponents of the Paretian revival in the 1930s—the central question for economists shifted to whether it was possible to characterize choice under conditions of risk and uncertainty in a manner analogous to that established for certainty. We will see in Chap. 4 that this fundamental step is linked to the theory of expected utility developed by John von Neumann and Oskar Morgenstern in the mid-1940s, but the question of choice in a risky and/or uncertain context was of course well known to earlier scholars, although not treated systematically. This chapter is devoted to some of the earliest attempts to reason about choice in uncertain contexts.

2.2 Pascal's Wager

The simplified formal model usually employed to represent a decision under conditions of uncertainty is that of a decision matrix. In such a matrix, the rows represent the possible actions of the individual agent and the columns represent the 'states of nature' that may condition the results of those actions, while the values indicated in the cells are the outcomes of each action in each possible state of nature. The choice of an action thus corresponds to multiple 'consequences'—as the outcomes of choices are often called—since the choice is assumed to be made without knowing which state of nature will occur. The fundamental aspect of the model is that the outcome is conditioned on something that does not depend on the individual's will, whether that something is a random mechanism, as in the case of buying a lottery ticket, or the realization of an event, as when the ticket refers to a horse participating in a horse race. What economists have really meant by the concept of the state of nature, particularly since the 1950s, is a problem that will be addressed in Chap. 5. But the intuition underlying the matrix approach can be traced back to the studies of a group of French philosopher-mathematicians in the seventeenth century, a period to which the first systematic developments in the field of probability theory are attributed.

The volume *The Logic, or the Art of Thinking*—published anonymously in 1662 and known as *The Logic of Port Royal* because it was attributed to a group of French

science alluded to in the introduction and illustrated in Chap. 4. The relevance of this distinction has subsequently been called into question. Starting with Herbert Simon's theory of bounded rationality, and then within the framework of so-called behavioral economics, a different approach was developed. We will return to these aspects, albeit only briefly, in Chap. 7.

scholars who met near the monastery of Port Royal, in the province of Paris—addressed the topic of deductive reasoning in the context of games of chance.[3] In Ian Hacking's (1975) unrivaled history of probability, the relevance of the *Logic* of Port Royal is evidenced by the fact that the volume was translated into all European languages and was still used as a textbook at Oxford in the nineteenth century. A number of other relevant works can be attributed to this period, such as Christiaan Huygens's *On Reasoning in Games of Chance*, which is regarded as the first accomplished textbook on probability. But Hacking points out that it is only in the last part of the *Logic*—attributed to Blaise Pascal's (1623–1662) influence—that the term probability was used in the modern sense of something that is susceptible to numerical measurement.

More importantly, in the *Logic*, probability has a dual character. Certainly, it applies to games of chance and the aspects of combinatorial calculus typically associated with them: probability in the strict meaning of chance. But the same term is also used as a notion to systematize reasoning in more general contexts, and is therefore, in principle, suitable for analyzing any uncertain situation. We will return to the meanings that can be attributed to the notion of probability, also from a philosophical point of view, in more detail later, but, as Hacking has pointed out, in the *Logic* it already appears in an articulated manner. On the one hand, there is the reference to 'aleatory' probability, that which relates to the stochastic laws of random processes, and which is usually identified with the category of 'classical' probability on the basis of the works of Jacques Bernoulli and Pierre Simon de Laplace. On the other hand, there is the reference to 'epistemic' probability, which refers to the degree of belief in a proposition or in the possible realization of an event, without any necessary reference to stochastic elements, and which would later be explored in the 'logical' probability approach of John Maynard Keynes and in the Ramsey-de Finetti-Savage 'subjective' probability approach.[4]

In the history of mathematics, Pascal is best known for his discussion of problems of pure chance. The birth of the mathematical theory of probability is usually traced back to an exchange with Pierre de Fermat, in which Pascal solved the problem of how to divide the winnings of a game of chance if the game is interrupted before it is

[3] Gambling or game *of chance* typically refers to recreational activities for profit in which winning or losing is predominantly random, where skill is not relevant, as in the game of dice.

[4] It must be reminded that the 'frequentist' approach to probability, which posits that the probability of an event is given by the frequency with which the event occurs over a long series of trials, would only manifest itself in a systematic manner at a later point in time, particularly following John Venn's presentation of this concept in his 1876 work, *The Logic of Chance*. It is also interesting to note that in his *Treatise on Probability*, which will be at the center of the next chapter, Keynes ([1921] 1973, p. 87) quoted a passage from *Port Royal's Logic* to recall that the subjective component of probability was central before the emergence of the frequentist approach, against which he intended to argue. The quote is as follows: "In order for me to judge of the truth of an event, and to be determined to believe it or not believe it, it is not necessary to consider it abstractly, and in itself, as we should consider a proposition in geometry; but it is necessary to pay attention to all the circumstances which accompany it, internal as well as external. I call internal circumstances those which belong to the fact itself, and external those which belong to the persons by whose testimony we are led to believe it."

completed. In this context, the probability of an event is logically deduced from the identification of the set of possible cases, understood as the outcomes of the game, and the determination of the number of cases in which the event occurs. If the cases are mutually exclusive, in the sense that if one occurs the other cannot occur, and if they exhaust all the possible realizations of the event, then the probability of an event E is given by the number of cases m in which E occurs with respect to the n possible cases, i.e., $p(E) = m/n$. This 'classical' view of probability was inspired by Pascal's analysis of games of chance and constituted the basis of Daniel Bernoulli's reasoning, which we will discuss in the next section.

It is usually Laplace's 1814 work *Philosophical Essays on Probability* that is credited with generalizing this approach to an epistemic view (Gillies, 2000).[5] Laplace integrated the idea of probability into the framework of the European Enlightenment, thus making it part of an approach that placed faith in the ability of science to provide explanations of the universe that could transcend religious arguments. With Laplace, probability came to be considered applicable to all situations in which knowledge of phenomena is actually ignorance, in the sense of lack of knowledge. If this ignorance concerns an environment that can be represented by a set of mutually exclusive and exhaustive cases, and these cases are all equally possible because there is no meaningful reason to believe that one will manifest instead of another, then each case can be assigned the same probability, i.e., if n are the possible cases, the probability of each case is $p = 1/n$. This method of assigning probability was called by Laplace the 'principle of insufficient reason.'[6]

But for a history of uncertainty in economics, Pascal is important for another analysis, in which he discussed a particular case, the so-called Pascal's wager. We are not interested here in the philosophical context of the argument, which probably relates to Pascal's refusal to follow the Jesuit approach, according to which any action that pursues the good is justified, regardless of whether there is a large or small probability of success.[7] What is relevant for our purposes is the structure of Pascal's argument, and in particular his ability to show that the possibility of thinking the

[5] The role of Thomas Bayes, who presented the first definition of subjective probability in a memory published in 1763, remained unrecognized for long (Stigler, 1982).

[6] From a mathematical point of view, probability is defined as a measure over a set S that satisfies certain properties. A probability is the function p that assigns to each subset A, B, … of S, a real number such that: (1) $p(A) \geq 0$ for each $A \in S$ (non-negativity); (2) if $A \cup B = \emptyset, p(A \cup B) = p(A) + p(B)$ (additivity); (3) $p(S) = 1$ (normalization). The mathematical basis of probability theory taught in university courses to this day found axiomatic justification in Kolmogorov's work in the 1930s, and is independent of his philosophical interpretations (Kolmogorov, [1933] 1950). But it remains inherent in the 'classical' approach.

[7] This is how Keynes ([1921] 1973, p. 340) interpreted it in his *Treatise on Probability*, where he argued: "A more correct doctrine was brought to light by the efforts of the philosophers of the Port Royal to expose the fallacies of [the Jesuit doctrine of] probabilism. 'In order to judge,' they say, 'of what we ought to do in order to obtain a good and to avoid an evil, it is necessary to consider not only the good and evil in themselves, but also the probability of their happening and not happening, and to regard geometrically the proportion which all these things have, taken together'." Keynes acknowledged that *Port Royal's Logic* was arguing in terms of what would later be called 'expected utility.'

	God exists	God does not exist
To be a believer	∞	x
Not to be a believer	$-\infty$	y

Fig. 2.1 Pascal's wager

uncertain need not be limited to games of chance, but could be applied in much more general contexts. Hence it was the 'art of conjecturing,' as Jacques Bernoulli would later to call it, a way of thinking that could be applied even outside a randomized context, up to cases where the probability is patently unknown.

Pascal analyzed the dilemma that a person faces when confronted with the question of the existence of God. Pascal's aim was not to present a religious argument or to justify an atheistic attitude, but simply to assess the meaning of the person's decision whether to believe or not. To do this, the necessity to consider both the utility of consequences and the probabilities of events was explicitly evidenced. On the one hand, either 'God exists' or 'God does not exist,' and there is no way to verify which of these two states is true; on the other hand, the person can choose either to believe in the existence of God, 'to be a believer,' or 'not to be a believer.'[8] In a schematic representation, each possible choice corresponds to a utility outcome, conditional on the existence or non-existence of God, as in the matrix in Fig. 2.1

If God exists, then being a believer leads to bliss, whereas not being a believer leads to eternal damnation, hence the infinitely positive and infinitely negative consequences. If God does not exist one can think that there is no difference between believing or not believing, hence $x = y = 0$. But one can also assume that, if God does not exist, being a believer may have led to renunciations that could have been avoided, in which case x takes on a negative value. Or one can assume that, if God does not exist, not believing may have led to the enjoyment of behavior not restricted by rules, generating a consequence y that takes on a positive value. In both cases, it would be necessary to consider the case where $y > x$.

It is important to note that, since Pascal's argument can be represented in a 2×2 matrix, it does not require the identification of a specific function associating consequences with choices, so it is not clear whether the outcomes he was talking about were expressed in terms of utility, rather than money. From the way he argued it seems more convenient to conclude—as Keynes suggested in his comment just quoted in footnote 7—that when Pascal used terms like gain and loss of bliss, he was thinking not of monetary gains, but of the utility associated with consequences. But the argument is important for the probabilistic aspect, regardless of the use or non-use of the utility of consequences, and so we focus only on that here.

---

[8] Hacking (1975, p. 71) points out that Pascal commented in handwriting: "The metaphysical proofs of God are so removed from the reasoning of mankind, and so complicated, that they have no force." And to emphasize the purely speculative character of reasoning, Hacking stresses that Pascal did not speak of 'being a believer' or not, but of 'acting as if one were a believer' or not. As far as we are concerned, however, the distinction is inessential.

Pascal developed his argument in successive steps. First, he examined the problem under the assumption that $x = y = 0$: if God does not exist, then there must be no difference in terms of consequences between believing and not believing. In this simplified case, Pascal concluded, it is better to believe: there is no need to question what the state of nature is, nor the probability with which the event 'God exists' versus the event 'God does not exist' occurs, because, regardless of whether God exists or not, it is always convenient to choose to believe. Pascal was applying what is known in modern decision theory as a (weak) dominance principle: if one choice always leads to a better (not worse) outcome than the other choice, then it dominates the other. But if the person who decides not to be a believer enjoys greater utility than the believer if God does not exist—so that one can assume that $y > x$—the logical argument based on dominance is inconclusive. This is where Pascal introduced the probability of the event occurring into the argument.

Pascal suggested that even in this second case, although there is no dominance of the results of one choice over the other, a person can proceed with the choice. She can do so on the basis of her expectations, which depend on the probabilities with which she thinks the state of nature can occur. Pascal first analyzed the possibility that the probabilities of the existence or non-existence of God are equivalent, and therefore equal to ½. This corresponds, as we have already mentioned, to what Laplace would later call the principle of insufficient reason, that is, the absence of evidence in favor of one hypothesis or the other. Here, Pascal noted that the mathematical expectation of choosing to believe is greater than that of not believing, whatever the finite sum y is.[9] The proposed criterion is therefore to choose the decision that maximizes the mathematical expectation of the consequences.

However, Pascal also noted that, given the consequences of the matrix, and in particular the bliss associated with being a believer if God exists, this holds true for any finite probability that God exists, however small. In fact:

$p(\infty) + (1 - p)x > p(-\infty) + (1 - p)y$, for any finite quantity of x and y, if $p > 0$.

Thus, even if the specific probability is not known, it is possible to reason about a range of possible values of the probability that justify the choice, which in this case is $p \in (0,1)$.

From the perspective of decision theory, it is noteworthy that with Pascal's wager subjective probability is used to extend the scope of the theory from certain outcomes to uncertain ones. In light of Pascal's argument, it becomes evident that decision-making under uncertainty is inherently probabilistic in nature: once the set of possible hypotheses about the outcome of an event has been defined, the application of probability calculus to this set can be undertaken. Moreover, the structure of the argument is constructed to investigate how one should act, not what one should think or believe:

[9] Here, as in the following, a mathematical expectation is defined as an average of values, each multiplied by the probability of its realization. Since by definition the sum of the probabilities of disjointed events covering the entire range of possible events is equal to 1, the mathematical expectation is a probability-weighted average of the possible values. This weighted average is also called the expected value.

the focus is on the decision to be made, while introspective references to motivations, whether religious or otherwise, are downplayed.

In conclusion, although an isolated example in a context of discussion that will continue to focus on risky situations such as games of chance, Pascal's argument is significant because it suggests that even the treatment of uncertainty can be approached with well-defined analytical methods.

2.3 The St. Petersburg Paradox

Among the debates of philosophers and mathematicians who sought to provide a mathematical interpretation to moral problems, anticipating the efforts of modern economists to systematize decisions under uncertainty, there is at least one other important episode worth reviewing. This episode has gone down in history as the 'St. Petersburg Paradox.' Although in Pascal's reasoning there was room for valuing the outcomes of choices in terms of their expected utility, the prevailing procedure for analyzing games of chance at the time was, and remained for a long time, that of considering the expected monetary gains. The conventional way of valuing choices in games was that to select the action with the highest expectation, that is, the action with the highest expected value—a method explained by Huygens in the aforementioned treatise. In a paper published in 1738 for the Imperial Academy of Sciences in St. Petersburg, Daniel Bernoulli (1700–1782), a member of a family of Swiss scholars of statistics and mathematics and the grandson of the aforementioned Jacques, advanced the argument that a more reasonable way to evaluate risky decisions existed.

Bernoulli ([1738] 1954) criticized the conventional way of the "measurement of risk" among mathematicians of the time, according to which a value equivalent to that of the bet (lottery) was given by a sum of money equal to the average of the winnings weighted by the probabilities with which they were likely to occur, that is, its expected value. Bernoulli offered a whole series of theoretical considerations to counter the employment of the mean of the probability-weighted monetary gains—or expected value—and proposed as an alternative to consider the weighted mean of the utilities—or expected utility—which he called "moral expectation" (or "mean utility," as translated from the Latin *emolumentum medium*).[10] In particular, he stated that the analysis of situations such as those involving lottery tickets had to be equated with that of goods in general, where the value of a good cannot be based on its price, but rather on the utility it yields to the individual. Since this utility is a subjective matter, "dependent on the particular circumstances of the person making

[10] Bernoulli's "fundamental rule" is as follows: "*If the utility of each possible profit expectation is multiplied by the number of ways in which it can occur, and we then divide the sum of these products by the total number of possible cases, a mean utility [moral expectation] will be obtained, and the profit which corresponds to this utility will equal the value of the risk in question*" (Bernoulli [1738] 1954, p. 24, original italics). As will be made explicit in Chap. 4, if we consider a lottery as represented by a set of possible outcomes, x, and associated known probabilities, p, then the lottery is valued according to $\Sigma_i \, p_i \, u(x_i)$.

the estimate" (Bernoulli, [1738] 1954, p. 24), when valuing a lottery, it is important to consider the initial wealth of an individual, and to distinguish between individuals with different initial wealth, because this typically influences the valuation of assets.

In this context, Bernoulli also proposed a specific function representing the utility that can be derived from the possession of assets or wealth in general. This was an increasing utility function characterized by downward concavity, hence with a positive marginal utility that decreases as the quantity possessed increases. Bernoulli represented total utility as a function of wealth, understood as the "quantity of goods already possessed." Assuming that "*the utility resulting from any small increase in wealth will be inversely proportionate to the quantity of goods previously possessed,*" Bernoulli ([1738] 1954, p. 25, original italics) showed that this function is the logarithm of wealth.[11]

Although the specific function he used appears to have been designed purely for calculation purposes, Bernoulli's assumption was that it is possible to measure both the total utility derived from wealth and the marginal utility derived from increases in wealth, and that marginal utility decreases as wealth increases. The figure from Bernoulli's English version of the original article—where total utility takes the shape of a logarithmic function (Fig. 2.2, with initial wealth, point B, attributed an initial utility value of A)—is basically the one still used today to represent risk aversion, a notion first formally introduced by Milton Friedman and Leonard Savage (1948). We will see in Chap. 4 that, after the axiomatic justification proposed by von Neumann and Morgenstern in the mid-1940s, the maximization of expected utility came to be

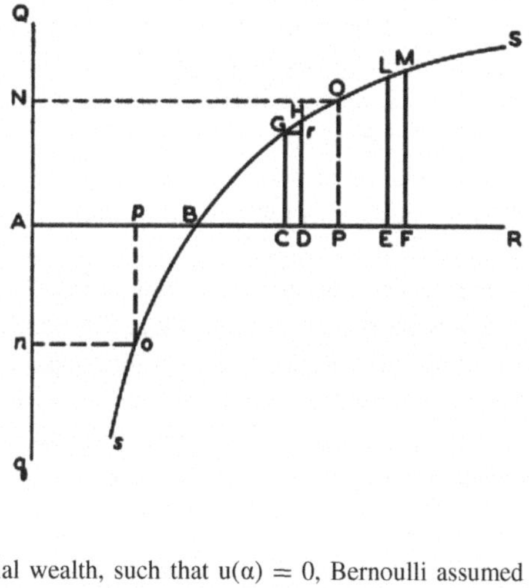

Fig. 2.2 Bernoulli's utility function (reproduced from Bernoulli, [1738] 1954, p. 26)

[11] If u is utility, x is wealth, and α is the initial wealth, such that $u(\alpha) = 0$, Bernoulli assumed $du = \beta(dx/x)$. Marginal utility would then be $du/dx = \beta/x$, with the parameter β representing a subjective element. Disregarding the constant, by integrating over the interval from α to x, $u = \beta log(x/\alpha)$. Under this assumption, the marginal utility of wealth is an equilateral hyperbola, whereby the way in which it decreases as wealth increases in percentage terms is constant: for example, a 10% increase in wealth is associated with a 10% decrease in marginal utility. It is noteworthy that this property is peculiar to functions satisfying the so-called absolute constant risk aversion of Arrow and Pratt (Pratt, 1964).

	HEADS	TAILS
KEEP THE TICKET	20,000	0
SELL THE TICKET	10,000	10,000

Fig. 2.3 Bernoulli's "poor man" choice problem

identified also as the 'Bernoulli proposition,' and the scholars who adopted it were called 'neo-Bernoullians.'

Bernoulli mixed his theoretical considerations with examples, on the basis of which he rejected the conventional proposition that a wager is worth its expected value. He examined the case of a "poor man" who happens to find a lottery ticket that pays 20,000 ducats with probability ½ and zero ducats with probability a ½, as if the outcome of the lottery depended on the toss of a fair coin. He then asked: "Will this man evaluate his chance of winning at ten thousand ducats?" (Bernoulli, [1738, 1954, p. 24).

In modern matrix form, the choice problem can be represented as in Fig. 2.3, again with a 2×2 matrix, but this time assuming that the probabilities of the event are known (½ for heads, ½ for tails).

According to the conventional interpretation since Huygens, a bet was considered fair if its expected value was equal to the price to buy it. Thus, 10,000 ducats were supposed to be the 'fair' price of this lottery ticket.[12]

Bernoulli, however, refused to believe that a poor man would follow the advice he could get from the theory of mathematical expectation of winnings, that is, to consider the choices 'keep the ticket' and 'sell the ticket' as equivalent. Bernoulli considered it obvious that the poor man would choose to sell, getting 10,000 ducats for sure. He then added that he thought it reasonable that this might happen even at 9000 ducats. Moreover, if the poor man were willing to sell the lottery ticket for 9000 ducats, he would probably find a rich person willing to buy it at that price. According to Bernoulli, as just mentioned, winnings can only be valued according to their utility, and utility is subjective.[13]

But what remains in history as Bernoulli's main argument against mathematical expectation is based on a real counterexample, the so-called St. Petersburg Paradox. After theoretically arguing for the necessity of evaluating lotteries on the basis of their expected utility, including through examples such as the one given above, and after showing how it was possible to calculate the sure sums he considered

[12] This is, of course, still common practice today. A bet is considered 'fair' if the price to buy it is equal to its expected value.

[13] We note here, but will return to this later, that the matrix representation can also be used to ascertain the value of the lottery ticket for a person. If we enter an unknown value in the second row, the same value for both columns, and ask the person to tell us what is the (minimum) value that would make the decision to keep the ticket equivalent to the decision to sell it, her reservation price for the lottery ticket can be obtained. Indeed, Bernoulli's argument was that the poor person's reservation price is lower than the rich person's.

equivalent in these examples on the basis of the specific utility function introduced, Bernoulli ([1738] 1954, p. 31) explicitly stated that he did not want to limit himself to considerations that could be mislabeled as "abstractions resting upon precarious hypotheses."

To this end, he introduced an example due to his cousin Nicholas Bernoulli, who had brought it to the attention of the mathematician Pierre de Montmort in an exchange published in 1713. As reported in the English translation of Bernoulli's paper, the problem is as follows:

> Peter tosses a coin and continues to do so until it should land 'heads' when it comes to the ground. He agrees to give Paul one ducat if he gets 'heads' on the very first throw, two ducats if he gets it on the second, four if on the third, eight if on the fourth, and so on, so that with each additional throw the number of ducats he must pay is doubled. Suppose we seek to determine the value of Paul's expectation. (Bernoulli, [1738] 1954, p. 31)

It is therefore a matter of determining the maximum amount of money that Paul should be willing to pay to participate in Peter's game, what in modern microeconomics textbooks would be called Paul's 'reservation price.' The paradoxical aspect arises from the calculation of the mathematical expectation of the monetary winnings, which was the value usually used in the discussions of the time: the mathematical expectation is given by the sum of a series of terms where the probability multiplies the winnings according to the rule $(½)^i 2^{i-1}$, with $i = 1, 2, \ldots n$, i.e., $(1/2)\,1 + (1/4)\,2 +$ $(1/8)\,4 \ldots$. Since each term is equal to 1/2, and since n goes to infinity, $\sum_i (1/2)^i 2^{i-1}$ is an infinitely high value, a sum that no Paul would pay to participate in this type of game. Moreover, the bet is fair: for every possible realization of the coin toss, in Bernoulli's example, the calculated value is by definition fair, so the price calculated as the sum of the fair values is itself fair.

Bernoulli argued that only by including the utility of the winnings in the argument was it possible to calculate a price that Paul could reasonably be expected to pay. Bernoulli calculated different prices for different values of the initial wealth and showed that the expected prices would increase as the initial wealth increases, but that they were all finite. In doing so, however, he was not really concerned with whether or not utility was a measurable quantity. He was simply applying an evaluation of monetary values using the logarithmic utility function, which he had introduced as practical for computational reasons. Thus Bernoulli concluded—essentially by calculating what is today known as the 'certainty equivalent' of a lottery—that this utility function showed that even with extremely high initial wealth it was impossible to imagine an unlimited willingness to pay.[14]

Before proceeding, it would be beneficial to offer a general reflection on the nature of a 'paradoxical' argument, defined as an argument that is supported by a counterexample. The issue of paradoxes of choice will resurface during the period of the establishment of expected utility maximization as the prevailing method of

[14] This argument is correct, but only in the presence of an upwardly bounded utility function, the mathematician Karl Menger ([1934] 1979) later observed. The definition of 'certainty equivalent' will be presented in detail in Chap. 4, but it serves to identify the sum of money whose utility is equal to the expected utility of a lottery, as is known from microeconomics textbooks.

choice under both risk and uncertainty. Bernoulli's seminal contribution to this field pertains to the specific case of the 'fair price' of a lottery being unreasonable to pay, regardless of initial wealth. While it is conceivable that an impoverished individual might, for some inscrutable reason, choose not to sell the lottery ticket from the initial example for 10,000 ducats, it appears totally implausible to envision anyone willing to pay an infinite sum—or even a substantial amount—to participate in St. Petersburg's game. The paradox lies in the fact that the prevailing theory, to which all theoreticians adhere, suggests a general solution that is ill-suited for a very clearly defined case. When confronted with this particular case, both the common sense of the decision-makers operating in reality—be they poor or rich—and an accurate, albeit at the time unconventional, theoretical investigation concur in denying the applicability of the traditional theory.

In the final part of his paper, Bernoulli reported the remarks made by the mathematician Gabriel Cramer to his cousin Nicholas, after noting that Cramer had also criticized his fellow mathematicians who accepted the criterion of the expected value of monetary gains. Commenting on the discrepancy between mathematical considerations and "popular valuations," Cramer had stated:

> I believe that it results from the fact that, in their theory, mathematicians evaluate money in proportion to its quantity while, in practice, people with common sense evaluate money in proportion to the utility they can obtain from it. I believe that this arises from the fact that, *in their theories*, mathematicians value money in proportion to its quantity whereas, *in practice*, people with common sense value money in proportion to the utility they can obtain from it. (Bernoulli, [1738] 1954, p. 33, original italics)

The question thus arises as to what to do with a theory that is manifestly not being followed as a guide—a norm of behavior—by those for whom it was conceived when they make real choices. This is especially true when the theory is not even considered convincing by some theorists, as was the case, as Bernoulli pointed out, with the famous mathematician Cramer. As will be discussed in Chap. 6, when Daniel Ellsberg proposed his paradox of choice in the early 1960s—which remains to this day the crucial critical argument against expected utility maximization as a criterion for choice under uncertainty—he noted the methodological analogy with the St. Petersburg Paradox, quoting Cramer's sentence just reported for its emphasis on the difference between "in their theories" and "in practice" (Ellsberg, [1962] 2001, pp. 48–49).

2.4 Utility Theory

The contributions of Pascal and Bernoulli are representative of an interest in logical reasoning in situations that are uncertain, or at least characterized by risk. Pascal introduced the possibility of using probability as a means to measure individuals' beliefs within the decision-making process. Bernoulli, meanwhile, identified the necessity to evaluate the expected values of utility instead of consequences (outcomes). However, economists largely disregarded these issues for an extended period. Consequently, when Frank Knight proposed his precise distinction between risk and uncertainty in 1921, it marked the first systematic examination of the subject, as will be discussed in the next chapter.

Indeed, the limited insights on the subject that can be found in Adam Smith, to begin with the Scottish scholar regarded as the founder of political economy, are not particularly well articulated. As later became customary in the early manuals of political economy, Smith discussed the phenomenon of risk outside of considerations of betting and gambling, focusing more on purely economic phenomena, such as the use of the resources possessed by an economic individual, be it the choice of what job to undertake or how to invest one's capital. In the chapter of his *Wealth of Nations* devoted to "Wages and Profit in the different Employments of Labour and Stock," risk is identified as one of the characteristics that explain the differences between worse- and better-paid occupations in terms of the "probability or improbability of success in them" (Smith, [1976] 1966, p. 117).

In his analysis, however, Smith did not take advantage of the clear resemblance to the phenomena of betting and insurance, nor did he aim at a systematic representation of them in the manner of Bernoulli. The rate of profit always increases more or less with the risk, but not enough to compensate for it completely, and indeed for undertaking certain risky investments the "presumptuous hope of success" seems to be the more relevant aspect, to the extent that competition in these fields often becomes excessive and leads to a reduction of profits "below what is sufficient to compensate the risk" (Smith, [1776] 1966, p. 128).

Even after Smith, the question of choice under conditions of risk and uncertainty was treated unsystematically, almost anecdotally, and Bernoulli's utilitarian approach was largely overlooked, at least until it was brought to the center of analysis by marginalist scholars in the 1870s. Indeed, the first economist to refer to Bernoulli's expected utility was William Stanley Jevons. With the marginalists, it is worth recalling, the theory of value broke away from the classical approach, for example that of David Ricardo, who had identified the main cause of the value of goods in the amount of labor required to produce them. Conversely, the value of goods became a question of the utility that can be derived from their possession, and in particular of the "degree of utility" associated with different quantities of a given good. This concept was refined by Jevons, who identified the "final" degree of utility, pertaining to the last unit of a good possessed, as "the all-important element in Economics" (Jevons, [1879] 2013, p. 147).

In dealing with marginal utility, Jevons stated as a "general law" that it decreases as the quantity of the good increases. Jevons ([1879] 2013, p. 159) noted that even before the birth of a proper science of political economy, mathematicians interested in games of chance had realized that they had to consider utility as well as probability, and that the utility of the same amount of money might be different for different people. In particular, he explicitly referred to Bernoulli and his idea of distinguishing the moral expectation from the mathematical expectation.

Jevons is regarded as one of the first scholars who emphasized the importance of the mathematical representation of the problems addressed by political economy. He argued that "the future progress of Economics as a strict science must greatly depend upon our acquiring more accurate notions of the variable quantities concerned in the theory," such as in his attempt to "express the laws of the variation of utility numerically" (Jevons, [1879] 2013, p. 146). But he was not inclined to employ the calculation method proposed by Bernoulli. Rather, he used Bernoulli's work as a means of invoking the authority of mathematicians to assert that marginal utility could only be generally decreasing. As for the results that derive from the application of moral expectation, Jevons noted that it was one thing to calculate the expected utility and quite another to take into account the pleasure derived from betting. Therefore, he concluded, that while it was well known that "gaming is, in the long run, a sure way to lose utility," pure economic theory had to acknowledge that gamblers were driven by motivations that fall outside its purview.

What Jevons meant by claiming that gambling inevitably results in a loss of utility was not made explicit in the original text. But this statement is indeed accurate when applied to the case of an individual with decreasing marginal utility who participates in a fair bet, as in Bernoulli's example of the poor man examined above.[15] Moreover, Jevons was aware that following Bernoulli's line of reasoning would prove "the advantage of insurance in all circumstances," which is indeed the case when insurance can be obtained through the payment of a fair premium.[16]

Jevons's stance on this matter was therefore ambivalent. Although he made reference to the concept of expected utility, he concluded that the phenomenon to which it was applied could find a convincing explanation mainly outside the realm economic theory. Indeed, although the bets on offer were typically unfair, gambling was extremely widespread in English society at the time. In Victorian England, it should be remembered, gambling was so diffuse to be consider a scourge of society, a vice that Jevons certainly did not intend to justify (and Marshall's remarks below can also be read in this light). However, after Jevons's analysis, economists could no longer avoid addressing the issue of risk in relation to utility.

In the period under consideration, that is, the late nineteenth century, the authority on the subject was undoubtedly Alfred Marshall, the systematizer of marginalist thought in the so-called neoclassical form. Marshall was both a political economist in the traditional sense and a profound expert of mathematics. He contributed a great deal to the introduction of the mathematical tool into economics, but concluded that

---

[15] Returning to the example of Bernoulli's poor man, if he prefers 10,000 for certain to a combination with probabilities ½ and ½ of 20,000 and 0, this means that $u(10{,}000) > \frac{1}{2}\,u(20{,}000) + \frac{1}{2}\,u(0)$. But since the left member can be rewritten as $\frac{1}{2}\,u(10{,}000) + \frac{1}{2}\,u(10{,}000)$, then the inequality can be read as $\frac{1}{2}\,[u(20{,}000) - u(10{,}000)] < \frac{1}{2}\,[u(10{,}000) - u(0)]$. This inequality is only satisfied if the increase in utility when going from 10,000 to 20,000 is less than when going from 0 to 10,000, which is implied by a less than proportional increase of total utility, that is, a decreasing marginal utility. Under these conditions, it cannot be rational to pay the expected value to buy a fair bet.

[16] By analogy with the definition of a fair bet, a fair insurance premium is one that is equal to the probability of realization of the event one is insuring against. For example, if the probability of a house fire is estimated to be 1%, it is fair to pay one cent per euro insured. As is well known from microeconomics textbooks, the fact that insurance at a fair premium is compatible with the maximization of expected utility is one of the first results obtained when using decreasing marginal utility to represent risk aversion (Varian, 2011, Ch. 12). It is notable that Bernoulli ([1738] 1954, pp. 29–30) had already established this result in his seminal contribution.

it should be used in a limited way. Marshall's idea of how the economist should go about formulating economic reasoning on a mathematical basis is well known: mathematics should not be used as an 'engine of inquiry,' but as a form of simplified language. Mathematics is useful for developing an argument that can then be translated back into ordinary language, and that proves its effectiveness for application to real cases, otherwise it should not be used (Dardi, 2016). Indeed, in his *Principles of Economics*—a 1890 volume which was the fundamental treatise on economic theory in the Anglo-Saxon world at least until its eighth edition in 1920—Marshall confined the mathematical argument supporting the considerations developed in the main text to a series of notes in the appendix, not least because the mathematics he used was not always elementary. In two of these notes from the mathematical appendix to his *Principles*, he discussed diminishing marginal utility as treated by Bernoulli.

Marshall ([1920] 1961) first referred to Bernoulli in the chapter where he introduced the concept of 'consumer surplus.' Starting from the representation of the demand function for a good, he showed in what sense the difference between the reservation price of consumers and the price they actually pay in competitive markets can be used as a measure of welfare, presenting the consumer surplus in the way it is still taught in microeconomic textbooks. In his commentary, he discussed the welfare that can result from an increase in income, and in this context, he mentioned Bernoulli's hypothesis. According to Bernoulli, Marshall noted, the utility associated with an increase in income depends on the initial level of income, and increases at a rate that is inversely related to income. In the mathematical notes devoted to Bernoulli, Marshall showed in detail that, as seen in the previous section, this is mathematically equivalent to assuming that the utility function is logarithmic. Marshall valued Bernoulli's hypothesis mainly because it could be used as a basic argument in the discussion of the principle of equal sacrifice in taxation: the hypothesis could justify the idea of taxing different subjects in such a way as to equalize the marginal losses in utility due to the imposition of the tax, an argument later used to justify the progressivity of income taxes.

Marshall ([1920] 1961, p. 111) then made a reference to the problem of betting, although only "in passing." From the fact that marginal utility is typically decreasing it can be deduced that "gambling involves an economic loss, even when conducted on perfectly fair and even terms," because the (negative) utility of the possible loss is greater than the (positive) utility of the possible gain. Marshall observed that, as he demonstrated in the second of his mathematical notes on the subject, any utility function with a positive, but at a decreasing rate, slope leads to this result. Similarly, it can be argued that a "theoretically fair insurance against risks is always an economic gain." In reality, however, insurance companies offer higher than fair prices, either to remunerate their capital or to cover themselves against the ongoing costs of possible losses from fraud; the question of insurance choice cannot therefore be treated in general terms, but "must be decided for each case on its own merits." And even with regard to gambling, Marshall added, the question cannot be resolved theoretically since the general principle proposed by Bernoulli is not sufficient to explain the excess of betting that is often observed. Such an activity can only be justified on the basis of the "feverish character" of gambling, which provides the gambler with an "excitement" that cannot be subjected to economic analysis (Marshall, [1920] 1961, p. 694).

In conclusion, we find in Marshall a more formal elaboration of the observations already made by Jevons, but the same conclusion: despite an endorsement of the pillar on which Bernoulli built his analysis, the principle of diminishing marginal utility, there is no consistent application of the criterion of choice identified by Bernoulli, that of maximizing expected utility. In particular, Marshall seemed to want to defend the proposition of diminishing marginal utility, indispensable to the marginalist economic theory of the time, against the observations of the actual individual behavior that might contradict it. Again, as with Jevons, there was no application of a general principle to explain choices involving risk. In addressing the problem of why some occupations were pursued despite being highly risky and generally not conducive to success, Marshall preferred to offer common sense considerations similar to those previously offered by Smith, rather than attempting to characterize the actions of a rational economic agent. The pursuit of an occupation that promises high rewards in only a few instances may be justified on the basis of an "adventurous disposition" or the attraction of the "social rank" it guarantees, rather than on the basis of an economic assessment of the "average good fortune" involved (Marshall, 1920, p. 461).[17]

As one of the leading historians of economic analysis, Mark Blaug (1997, p. 318), has noted, following Jevons's argument about the impossibility of applying expected utility maximization to choices concerning uncertainty, Marshall seemed to impose a real "ban" on its use, a ban that lasted until von Neumann and Morgenstern's analysis. The observation that it was common for real economic agents both to participate in clearly unfair lotteries and not to insure themselves at fair insurance premia, was considered in contradiction with Bernoulli's decision criterion.[18] In their influential assessment of the actual use of utility in economics prior to von Neumann and Morgenstern, Friedman and Savage (1948, p. 281) noted that the reason for the rejection of expected utility soon proved unjustified: the ordinalist interpretation provided by Pareto showed that the principle of decreasing marginal utility was not necessary for the theory of behavior under certainty. But they also pointed out that even in the late 1930s, on the same assumption as Marshall, an economist of the stature of Jacob Marschak considered the maximization of expected utility to be clearly "unrealistic," without realizing the character of generality it might assume.

[17] Another emblematic case is that of Knut Wicksell, the Swedish economist who was highly influential among economists in Continental Europe at the turn of the nineteenth and twentieth centuries. In his *Lectures on Political Economy*, Wicksell ([1902] 1977, p. 211) did not deviate from what we have already seen in Marshall, arguing that "in accordance with the general law of marginal utility" it is necessary to consider that the loss of a given sum of money has a greater utility (in absolute value) than the gain of the same sum. Therefore, when discussing investment activity and the fact that profit is typically uncertain, only a probability of success far greater than the probability of failure can justify the decision to invest. Wicksell considered uncertainty as a factor hindering the accumulation of capital, the effects of which could be mitigated by risk diversification and insurance. However, he did not undertake a systematic analysis of uncertainty either.

[18] On the nuances of the rejection of Bernoulli's expected utility by marginalist economists, see also Giocoli (1998).

We will see in Chap. 4 in what sense the von Neumann and Morgenstern approach rehabilitated the principle of expected utility maximization to the point of making it the pillar on which all modern economic theory bases the theory of decision-making under risk and uncertainty.

2.5 Concluding Remarks

In order to emphasize the importance of a theory of decision-making under conditions of risk that, like that under certainty, is based on an analysis of the utility of choices, Friedman and Savage (1948) argued that when economic agents make their choices, whether they realize it or not, there is always a risk involved. Whether they are choosing among possible occupations, among different investment activities, or among a set of financial securities, the choices economic agents have to make are somewhat analogous to those they make when deciding whether or not to place a bet. Economists have, of course, analyzed all these issues in general terms, but until the 1940s, as Friedman and Savage noted, there was no real integration between the way they analyzed choices under certainty and those under risk, and the explanation based on utility maximization remained confined to the former domain.

In this chapter, we have seen that probability and utility were analyzed by some philosopher mathematicians long before economists. We have also seen that economists accepted the approach suggested by these philosopher mathematicians, but they were reluctant to employ it as a general solution for decision-making, even after the marginalist revolution had identified a central role for utility. It is therefore to the contribution of von Neumann and Morgenstern that we must turn for the definitive appreciation of expected utility.

However, before illustrating the bases on which the principle of expected utility maximization was eventually accepted, in the next chapter we will present the reasons underlying the conscious rejection of this further step associated with the figures of Knight and Keynes. We will see that the perplexity about the approach that later became virtually dominant—if not its outright rejection—has much to do with the meaning attributed to the notion of probability, that is, whether or not it has the dual character attributed to it by Pascal. But we will also see that, underlying the interpretation of probability, is the more general question of how far economics can go in the formal mathematical representation of economic phenomena.

References

Bernoulli, D. ([1738] 1954). Exposition of a new theory on the measurement of risk. *Econometrica, 22*, 22–36.

Blaug, M. (1997). *Economic theory in retrospect* (5th ed.). Cambridge University Press.

Dardi, M. (2016). Philosophy and psychology of mathematics according to Alfred Marshall. *Cambridge Journal of Economics, 40*, 283–308.

Ellsberg, D. ([1962] 2001). *Risk, ambiguity and decision.* Garland Publishing.

Friedman, M., & Savage, L. J. (1948). The utility analysis of choices involving risk. *Journal of Political Economy, 56*, 279–304.

Gillies, D. (2000). *Philosophical theories of probabilities.* Routledge.

Giocoli, N. (1998). The 'true' hypothesis of Daniel Bernoulli: What did the marginalists really know? *History of Economic Ideas, 6*, 7–43.

Hacking, I. (1975). *The emergence of probability.* Cambridge University Press.

Jevons, W. S. ([1879] 2013). *The theory of political economy.* Macmillan.

Keynes, J. M. ([1921] 1973). *A treatise on probability. The collected writings of John Maynard Keynes* (Vol. VIII). Macmillan.

Kolmogorov, A. N. ([1933] 1950). *Foundations of the theory of probability.* Chelsea Publishing.

Marshall, A. ([1920] 1961). *Principles of economics* (9th [variorum] ed.). Macmillan.

Menger, K. ([1934] 1979). The role of uncertainty in economics. In Id. *Selected papers in logic and foundations, didactics, economics.* Reidel Publishing Company.

Moscati, I. (2018). *Measuring utility: From the marginal revolution to behavioral economics.* Oxford University Press.

Pareto, V. (1896). *Cours d'Économie Politique* (Vol. 1). Rouge.

Pratt, J. W. (1964). Risk aversion in the small and in the large. *Econometrica, 32*, 122–136.

Robbins, L. (1935). *An essay on the nature and significance of economic science* (2nd ed.). Macmillan.

Smith, A. ([1776] 1976). *An inquiry into the nature and causes of the wealth of nations. The Glasgow edition of the works and correspondence of Adam Smith* (Vol. II). Oxford University Press.

Stigler, S. M. (1982). Thomas Bayes's Bayesian inference. *Journal of the Royal Statistical Society. Series A, 145*, 250−258.

Varian, H. (2011). *Intermediate microeconomics.* Norton.

Wicksell, K. (1901 [1977]). *Lectures in political economy* (Vol. I). Augustus M. Kelley.

Chapter 3
Uncertainty According to Knight and Keynes

3.1 Introduction

In a 1950s textbook still recommended today as an introduction to game theory, the two mathematical economists Duncan Luce and Howard Raiffa distinguished risk from uncertainty as follows. The decisions of economic agents, whether individual or group decisions, can be classified as decisions that are made under conditions of certainty, risk and uncertainty. One speaks of: (a) certainty, when each decision invariably leads to a single specific outcome; (b) risk, when each decision leads to one of a set of possible specific outcomes, and when each outcome may occur with a given probability that is "assumed to be known to the decision-maker;" (c) uncertainty, when each decision leads to one of a set of possible specific outcomes, but the probabilities with which each outcome may occur are "completely unknown or are not even meaningful" (Luce & Raiffa, 1957, p. 13).

This distinction was used intuitively in the previous chapter and remains canonical among economists. Similarly, Itzhak Gilboa (2009, p. 3), one of the leading contemporary decision theorists, proposes to distinguish between decision domains according to how intuitively understandable the probabilities of events seem, in "increasing order of conceptual difficulty:" (1) you are about to toss a coin, what is the probability that it will come up heads? (2) you are about to park your car in the street, what is the probability that it will be stolen overnight? (3) you are about to undergo surgery, what is the probability that you will survive the medical operation? (4) you are about to start a new business venture, what is the probability that the financial markets will collapse by the end of the year? Clearly, it is gradually more difficult to understand the probabilistic context in which the decision has to be made. From a circumscribed and well-defined domain, such as that in case 1, one moves to increasingly complicated domains, up to those with extremely blurred boundaries,

C. Zappia, *Uncertainty in Economics*, Springer Studies in the History of Economic Thought, https://doi.org/10.1007/978-3-031-91221-4_3

such as that in case 4, where it appears difficult to formulate probabilities, because they may be regarded as unknown or meaningless.[1]

Probability therefore plays a crucial role in clarifying the distinction between risk and uncertainty. But probability itself is a concept susceptible to interpretation, with relevant differences between different interpretations. Indeed, as we will begin to see in this chapter, there are several 'philosophies' of probability, regardless of its characterization as a function with precise mathematical properties. To a first approximation, we can say that known probabilities refer to their 'objective' measurement, but in this chapter we emphasize the necessity for economists to assess the representability and measurability also of probabilistic judgments, called degrees of belief, from which a notion of 'subjective' probability emerges. The unknown probabilities of point c) of Luce and Raiffa's definition, or those with the highest degree of difficulty in terms of intuitive comprehension, as in Gilboa's point 4), evidently have the character of subjectivity. This does not mean, however, that they can remain outside the scope of analysis, especially in the field of economics, since even events to which it is difficult, apparently impossible, to assign a probability can be highly relevant to the results of a decision.[2]

As illustrated in the previous chapter, risk was clearly the theme addressed by Daniel Bernoulli and by most studies on gambling and betting, thus with probabilities given 'objectively,' in a context which, following Ian Hacking, we have called 'aleatory.' But in a paradigmatic case such as that imagined by Blaise Pascal, the idea was to apply a probabilistic approach even to issues that fall outside clearly defined contexts, up to and including extremely uncertain situations such as the existence of a superior entity, with probabilities that, even if one were to attempt to assign meaning to them, were at most 'subjectively' given. We have also observed that the early marginal economists did not dedicate sufficient attention to this topic, offering only a few and somewhat unsystematic considerations, as evidenced by the work of William Stanley Jevons and Alfred Marshall. Moreover, they represented a tradition of thought that originated with Adam Smith and was confirmed even later.

---

[1] Actually, in the cited work, Gilboa, an Israeli mathematical economist, proposes as case 4) "the probability of war in the Middle East by this year." A similarly suggestive but different example has been used in order to avoid references to the dramatic situation still existing at the beginning of 2025.

[2] A reference to the 2020 pandemic may be instructive. An investment activity that was planned in a normal (pre-pandemic) period and then has to be carried out when all economic systems are at a standstill due to the spread of the pandemic, may well represent a decision whose economic outcome is extremely uncertain to evaluate. Indeed, it depends on the occurrence of an event whose probability is unknown or to which it is difficult to attribute any significance. What do we make of the opinion of an expert who states that the scientific community had been wondering for some time about the probability of a pandemic event to occur, given that there are many precedents in history (the Spanish flu of 1918 or SARS in 2003)? What was the probability of a new pandemic in the year before it actually occurred? What meaning could be attached to a precise numerical indication of that probability? Did experts agree on the same numerical probability or did their opinions differ? How to take these aspects into account when deciding whether to invest in an increase in production capacity—to use an analogy with what, as we shall see, was the concern of Knight and Keynes—is obviously a crucial question in proposing a theory of decision under uncertainty.

In the history of political economy, the question of whether there is an important difference between risk and uncertainty was first explicitly and systematically addressed by Frank Knight. Knight employed the criterion of measurability in probabilistic terms to classify phenomena as either risky or uncertain. In the same period, the question of the measurability or otherwise of probabilistic phenomena was also addressed by John Maynard Keynes. In defining his philosophy of probability, Keynes questioned the significance of the numerical representability of probability. This chapter is devoted to an examination of the contributions of these two prominent figures.

3.2 Risk and Uncertainty in Knight

The American economist Frank Knight (1885–1972) was one of the founders of the so-called Chicago School of economics, a school that is still extremely influential in the global landscape of economic theory and policy. Belonging to the second generation of economists at the University of Chicago, Knight shared with Jacob Viner the merit of having transformed the Department of Political Economy of that university—founded in 1892 on the radical instances of institutionalists such as Thorsten Veblen—into a stronghold of neoclassical marginalism and, later, monetarist thought.

A native of a small town in Illinois, after graduating from the University of Tennessee Knight began his doctoral studies at Cornell University—at the time a bastion of neoclassical economic thought under the influence of Allyn Young—obtaining his doctoral degree with a study of the notion of profit in economics in 1916. His dissertation, which would later become his famous volume *Risk, Uncertainty and Profit*, earned him first an appointment at the University of Utah and then a call to Chicago in 1927. In Knight there was no apologia for the market mechanism as in later representatives of the School, such as Milton Friedman and Robert Lucas, but a defense of the core of neoclassical thought, which had not yet become orthodoxy in the United States. While not embracing the theoretical-mathematical approach of Léon Walras's general equilibrium theory, in the 1920s Knight set out to counter the empiricism inherent in much American economic thought at the turn of the century. To this end, he identified the characteristics of a system of perfect competition as the theoretical nucleus necessary, albeit abstract, for understanding any economic phenomenon. Knight was indeed an advocate of the importance of market capitalism, but without wishing to offer an all-out defense of it on an ethical level. Rather, he was interested in identifying its propulsive role, the unique capacity of such an organization of exchanges to generate opportunities for individuals.[3]

---

[3] From this perspective, Knight shared with Austrian economists, especially Friedrich von Hayek, a specific interpretation of the workings of competitive markets in capitalist societies. Both economists emphasized the characteristics of the competitive process as a mean of achieving equilibrium, rather than focusing on the end state of general equilibrium. They perceived the process to be contingent upon the role played by unknown information: imperfect knowledge in the case of Knight, specific and diffuse knowledge among economic agents in the case of Hayek (1937). While Hayek's analysis

However, the system of perfect competition, of which neoclassical price theory is the theoretical foundation, is an environment characterized by perfect knowledge of the data on which to base decisions. This is substantially a static approach, where there is no change and the economic system is analyzed as if "free from all disturbing influences." It is in this context, Knight noted, that it is grounded the ambition of economic science to be considered as an exact science, whose results depend on abstract reasoning, and not on the observation of reality and its alleged empirical regularities on which the American institutionalist economists of the time concentrated.[4] The results obtained under perfect competition provide an approximation of reality because they are believed to identify the situations toward which the system tends, and which would occur if the ideal conditions on the basis of which the theoretical reasoning is developed were met. But the simplification of reality required to develop reasoning under conditions of perfect competition is essentially an "assumption of practical omniscience on the part of every member of the competitive system" (Knight, [1921] 1964, p. 197). Since one cannot stop at this state of analysis, Knight argued, it is necessary to clarify economists' ideas about the nature and limits of knowledge, and the relationship between knowledge and behavior.

In a dynamic setting, a deterministic approach based on the use of unrealistic postulates such as perfect knowledge of the data by the agents cannot be applied. The state of perfect competition only serves to identify clearly the effects of a change, similarly to what happens in physics when a force is studied for its impact on a system in static equilibrium. With this in mind, Knight focused on the role of the entrepreneur as the engine of economic development, a role substantially neglected by the orthodox theoretical framework of the time.[5] According to Knight, in static economics, it can be observed that the effect of competition is the tendency for profits to be nil, and therefore the condition of normality of a theoretical system under perfect competition is zero profit. In fact—as it is still conventional in economics today—the profit component due to the normal remuneration of entrepreneurial activity is essentially considered a cost component like that of the other factors of production, including interest as remuneration of invested capital. The existence of 'pure' profit (sometimes called extra-profit) can therefore only be justified in a situation of 'real' competition, in which the entrepreneur confronts change. But this means to operate

of diffuse knowledge has been lauded for its accuracy and its role as a foundational model for the study of complex economic systems (Birner, 2015), he primarily focused on informational issues, without explicitly addressing the role of uncertainty in economics. Nevertheless, the influence of Hayek's economic thought can be discerned in the insights of George Shackle (McCann, 1994, pp. 85–86), a topic that will be examined in Chap. 7.

[4] The American institutionalist school, which exerted considerable influence at the time, was notable for its rejection of the idea that human behavior could be analyzed in an abstract manner, without consideration of the human habits derived from the cultural and social contexts in which the individuals being analyzed by economists operated.

[5] A notable exception was another Austrian economist of the time, Joseph Schumpeter, who shared with Knight a particular interest in the theory of economic development, although he did not explicitly deal with uncertainty.

under conditions of uncertainty, not those of certainty and sure outcomes that characterize the theoretical model of competition. It is therefore to uncertainty, Knight argued, that one must look in order to understand the existence of pure profit: this is justified only when investment activity takes place in a context of uncertainty.

Knight recalled that at least as far back as Smith, economists had recognized that profit must contain a component that is not simply a return on invested capital. But the connection between uncertainty and profit had no precise theoretical representation before Knight, who pointed out that this was due to the fact that economists, while noting the idiosyncratic character of the "human attitude toward uncertainty" of all kinds, had never really investigated the ambiguity implicit in an inaccurate use of the terms risk and uncertainty. To proceed in analyzing behavior, Knight argued, one must abandon the common usage of inaccurately referring to risk as an unfavorable contingency (the risk of a loss) and uncertainty as a favorable contingency (the uncertainty of a gain). For correct reasoning, it is necessary to denote by risk every situation of "measurable uncertainty," in which case one can use an objective numerical probability of the events that condition the results of choices; one must instead denote by uncertainty every situation of "unmeasurable uncertainty," in which case one must ask oneself what a subjective probability might mean. Only based on this assumption, that is, by making it clear in which cases it is possible or not to measure uncertainty by means of probability, is it possible to develop an analysis of the "rational grounds of action" (Knight, [1921] 1964, p. 236).

To the specific topic of decision under uncertain Knight devoted two chapters of his work, designed to broaden the analysis from perfect competition to actual competition. First, he attempted to precisely delineate the characteristics of the phenomenon of choice, proposing a taxonomy of risky and uncertain situations. Then he proposed to specify what the criterion is for dealing with uncertainty as distinct from risk. The first of the two chapters is almost methodological in nature, and it is here that Knight proposed the distinction between risk and uncertainty that is still customary in economics today.

As mentioned, Knight set out to clarify the relationship between assumptions about knowledge and economic behavior. The economic behavior normally studied in economics is based on an "exact" theory of knowledge, for which the deductive reasoning inherent in logic is applied. But the ability to understand the context in which one operates is in reality often limited. Most "real decisions," Knight argued ([1921] 1964, p. 214), are based not on precise deductive logic, but on "reasoning … of more tenuous and uncertain character," depending on the knowledge actually available. In order to better understand the basis on which decisions are made, only an analysis of the probabilities involved can properly investigate cases where there is no certainty.

In fact, Knight made it clear, under particular conditions a risky situation does not require any analytical tool other than those of certainty. Even if the data relating to the decision-making context are not known with certainty, but a "numerical probability" of its occurrence can still be assigned to each possible outcome, it is possible to study the behavior of agents precisely. Knight gave as an example the business of producing champagne, which is affected by the possibility of some bottles bursting before going into the market. This is a problem that entrepreneurs can assess correctly by including in their calculation of costs the proportion, which they know from experience, of the bottles that will burst on average in a production cycle. And even if the individual producers may find it difficult to identify this value, they can insure themselves against this eventuality according to a typical risk-sharing process, that is, either by sharing the risk with a group of other producers operating under similar conditions, or by purchasing insurance from an insurance company.

In general terms, therefore, the question is what is meant by probability and under what conditions a numerical probability can be assigned to events that condition the results of choices. Knight ([1921] 1964, pp. 224–225) distinguished between three different types of probability situations: (1) "a priori probability," when it is possible to classify the possible cases precisely, such as when one asks what is the probability that a roll of a perfect die gives the number 3, and since there are six possible outcomes of the roll of the die, the probability that the die gives 3 is 1/6, regardless of what the outcome of the individual roll is if one observes it; (2) "statistical probability," when it is possible to make an empirical evaluation of the frequency of a given outcome over a set of observations, a set that constitutes a homogeneous group of cases, such as when observing how many bottles of champagne in a production process explode compared to the total number of bottles produced; (3) "estimates," when the observations one makes cannot be evaluated in relation to a homogeneous group of observations, that is, when "there is *no valid basis of any kind* for classifying instances."[6]

Knight acknowledged that defining probabilities can also be a complicated matter in the first two cases, because economic reality almost never conforms to a perfect random mechanism, as assumed in the first case; nor can statistical analysis always be conducted by taking as reference a universe of cases that are all precisely analogous, as assumed in the second case. But in both the first and the second case one can abstract from these complications, and assume that the distribution of possible outcomes in a group of alternatives is known precisely. In this case, the probabilities are objective and the situation is a risky situation. Conversely, if the situation being discussed has a high degree of uniqueness—therefore precluding the possibility to refer to a group of cases against which to consider the individual case—there can be no objective probabilities, only "estimates." This situation is one of uncertainty.

We mentioned earlier, and we shall see in detail with both Keynes and later authors, that it is accepted that there may be degrees of belief interpreted as true probabilities, in this case intended as subjective probabilities. But Knight ([1921] 1964, p. 223) clearly indicated that the typical entrepreneurial economic activity consists largely in acting on estimates rather than on probabilities inferred from objective data, on "judgments" or "intuitions" rather than on rational reasoning in the strict sense. And although judgments and intuitions may mean something analogous to a "probabilistic

[6] It should be noted that, unlike Keynes, Knight did not explicitly address probability in his book. Consequently, he did not delve into the intricacies of the various approaches. We will therefore leave until the next section some considerations on the 'classical' and 'frequentist' views of probability, which coincide with Knight's probabilities of cases 1 and 2.

judgment," this can only be decidedly different from the objective probability of cases 1 and 2 above. Uncertainty would be eliminable if a set of cases analogous to the one under consideration could be imagined, but this is impossible, and in this sense uncertainty is "unmeasurable," because the conditions for the existence of a true probability do not exist.

Knight's specific contribution has been to highlight the role of this "third species under that genus [of probability]," which was substantially neglected by the economic theory preceding him. Uncertainty that is somehow reducible to an objective measure, a "quantitatively determinate probability" is not really uncertainty, and therefore not a problem for the entrepreneur making decisions, precisely because it can be assimilated to risk and therefore dealt with by multiplying the probabilities of the events that affect profits by the profits associated with each event. "*True uncertainty*" is a higher form of uncertainty, in the sense of not being susceptible to measurement, and hence to elimination. Only the entrepreneur who confronts this type of uncertainty can enjoy, if successful, the "peculiar income" that is called profit, because only true uncertainty prevents the "theoretically perfect outworking of the tendencies of competition" (Knight, [1921] 1964, p. 232).

According to Knight, the typical example of this is the entrepreneur's decision to enlarge the scale of his production process through a significant investment in production capacity. In planning this, the entrepreneur may consider all factors susceptible to measurement—both the future demand for goods he intends to satisfy and the future results of the actions he takes to satisfy that demand—but he will still end up with an end result that is an estimate of the probable outcome of his choices. However much one may delude oneself into giving probabilistic content to a choice of this kind, its essential characteristic is that it is a rare decision, essentially unique, not traceable to a set of similar decisions with respect to which the real probability of success of the case in question can be identified.

As we shall see below, a crucial point in all subsequent developments in decision theory has been the question of whether or not Knight's estimates can be regarded as subjective probabilities in the strict sense, that is, whether or not the mathematical properties of a probability function can be attributed to the estimates of the Knightian entrepreneur. On the one hand, Knight acknowledged that it cannot be denied that, in reality, entrepreneurs must necessarily make probabilistic judgments if they want to act, and that these judgments are often formulated as if they were true probabilities. Moreover, although entrepreneurs are often forced to make decisions that have a unique character, they would still try to refer to a group of similar situations. Knight ([1921] 1964, p. 227) philosophically reminded the reader that "nothing in the universe of experience is absolutely unique any more than any two things are absolutely alike." These considerations might suggest that Knight was after all talking about a kind of subjective probability, and the conclusion of some interpreters was indeed that, in the light of what was argued a few years later by Frank Ramsey and Bruno de Finetti on how to define such probabilities numerically, Knight's perplexities could be overcome (Arrow, 1951; LeRoy & Singell, 1987).

However, Knight made it clear that, even if it were possible to reduce estimates to subjective probabilities, the issue would still arise differently from risky situations with objective probabilities. In the case of the entrepreneur who has to make a crucial decision, there is the problem of the "confidence" with which one expresses a definite opinion, and the degree of confidence one places in an opinion is based on a logical or psychological reasoning that is "obscure, a part of the scientifically unfathomable mystery of life and mind" (Knight, [1921] 1964, p. 227). Knightian uncertainty is thus not representable in clearly defined probabilistic terms, that is, it is irreducible to risk, to use a formula employed by those interpreters who argue that it cannot be treated with the standard tools of prevailing decision theory (Langlois, 1994).

As already mentioned, Knight devoted a second chapter to the topic of uncertainty, focusing on how to deal with uncertain situations. He did not propose a specific decision-making criterion, which in any case he took for granted to be the calculation of the expected profit for each possible decision, when all possible alternatives are known and the probabilities of the realization of each one accurately calculated.[7] Instead, his analysis focused on how to reduce the influence of uncertainty on choices and in particular on risk insurance. It is here that Knight presented the other way in which the distinction between risk and uncertainty is conventionally referred to: for Knight, there are insurable risks and uninsurable risks, and uncertainty is an uninsurable risk.

In particular, the insurance market is able to cover risky situations because the insurance company can reason about the grouping of similar risks. By analogy with general reasoning, the insurance firm is able to calculate the probabilities of those events for which it has information on many similar occurrences, as in the case of life insurance, whereby the probabilities of death of two healthy individuals who are similar in terms of characteristics such as gender, age, occupation and lifestyle habits, are essentially the same. In contrast, this assimilation of cases cannot take place for uncertain situations: insurance against a specific kind of illness or accident is so conditioned by the particular conditions of the individual case that it precludes the possibility to rely on a grouping of cases.

It is well known, Knight admitted, that even in such peculiar cases insurance policies offered, but they are sold at a much higher price than they should be, precisely because the insurance company cannot treat them as a risk. The fact that some underwriters insure any kind of risk—as the Lloyd's of London are said to do when they agree to insure an economic activity against, for example, the Loch Ness monster being caught by a given date—only shows that these underwriters are convinced that they stand to gain from being able to demand extremely high insurance premiums that are in fact based on "almost pure guesswork," rather than on statistically calculated data (Knight, [1921] 1964, p. 250).[8]

[7] Although without placing any particular emphasis on it, Knight ([1921] 1964, pp. 198–199) essentially accepted the method of expected utility maximization: "Even though the businessman could not know in advance the results of individual ventures, he could operate and base his competitive offers upon accurate foreknowledge of the future if quantitative knowledge of the probability of every possible outcome can be had."

[8] The example of the Loch Ness monster is not Knight's, but it is often referred to in the literature on insurance of peculiar risks following the actual issuance of such a policy in 1971, when a famous Scottish whiskey brand, having promised a large sum of money to anyone who would catch the

Before moving on, it must be noted that, after having indicated such a clear and straightforward criterion for classifying risk and uncertainty, Knight did not return to the subject in his later studies. Although the 1921 volume was reprinted several times and Knight wrote new prefaces for these reprints, he did not comment on other ways of understanding risk and uncertainty, most notably Keynes's, not even to show appreciation for what seems to many to be at least a consonance of intentions. Arguably, the neoclassical approach to which he remained faithful prevented him from appreciating the novelties of the *General Theory*, of which he was a critic. The Knightian notion of uncertainty has therefore remained crystallized in its initial meaning.[9]

If probability is objectively given, uncertainty is measurable, hence it is a risk, intended as a situation where the probability of the individual case coming true is known, because it is derived either by a priori logical reasoning or by a statistic about the frequency of cases. True uncertainty, on the other hand, is a situation where this cannot happen due to the uniqueness of the individual case, that is, the impossibility of relating it to a group of cases in a meaningful way, so that what guides the individuals' behavior is an opinion rather than specific probabilistic knowledge. Since probability is not objectively given, uncertainty is not measurable, and although even in this case economic agents have to act, they do so on the basis of estimates, not true probabilities.

3.3 Keynes and Probability Theory

We have already pointed out in the introduction to this volume that the figure of John Maynard Keynes (1883–1947) is pivotal to a history of uncertainty in economics. Keynes devoted fundamental sections of his *General Theory of Employment, Interest and Money* to uncertainty, and in the important essay summarizing the themes of his *General Theory*—a text we already mentioned in the introductory chapter—he underscored uncertainty as the primary component of his analysis. Keynes (1937) contended that the underemphasized role attributed to uncertainty was the reason why the economic theory of the period failed to accurately interpret the phenomenon of unemployment.

From this point of view there is a clear assonance with Knight's argument, specifically when Knight argued that fundamental aspects of the functioning of the economic system such as investment activity, and the profit associated with it, can only be adequately addressed in a context where the knowledge attributed to economic agents

monster by a given date, turned to the Lloyd's underwriters to insure against such an event, and obtained the required insurance cover (Borch, 1976).

[9] Fiorito (2016) suggests that this may have been due to the evolution of Knight's thought. After *Risk, Uncertainty, and Profit* he became less confident about the scientific status of economics and more concerned about the nature of the knowledge it could provide. The gap between the idealized postulates of economic theory and the concrete reality of human experience in the face of uncertainty emphasized in 1921, was only a first step in this process.

is not perfect. The role of uncertainty in Keynes's economic theory, however, will be discussed in the next section of this chapter. Instead, the present section is devoted to illustrating a phase of Keynes's thought that long predates that of the *General Theory*, which, it must be recalled, was published in 1936.

Keynes grew up in Cambridge, the cradle of British political economy, which was the fulcrum of theoretical developments at the end of the nineteenth century, both in Britain and on the global stage. Despite the growing trend toward a more mathematical approach, as evidenced by Alfred Marshall's mathematical notes mentioned in the previous chapter, Cambridge economics was still deeply imbued with philosophical and moral discussions. The young Keynes found himself immersed in these discussions, both because he was the son of the prominent logician and moral science lecturer John Neville Keynes—who helped delineate the economic methodology of the time with his work *Purpose and Method of Political Economy*—and because he frequented interdisciplinary circles such as the *Apostles*, a secret society similar to a discussion group where he met, among others, the philosopher George Moore and the mathematician Bertrand Russell. An honors student in mathematics at Eton High School, Keynes opted to pursue his studies in mathematics on his way to university, gaining admission to King's College, Cambridge, in 1902.

After graduation, he decided to join the British civil service, serving as junior clerk in the Ministry of Indian Affairs, but continued his studies with the aim of securing a position as a university lecturer. Notwithstanding the influence of Marshall—a friend of his father's, who had supervised him in economics in preparation for the competition to enter the Indian Office—Keynes submitted a dissertation on the theory of probability, still thinking of a possible lectureship in mathematics at King's College. However, while revising his dissertation prior to its final approval in 1908, he accepted Marshall's offer to commence lecturing economics, later becoming a Fellow of King's College as an economist.

Extremely active even outside the strictly academic sphere, Keynes became internationally famous for *The Economic Consequences of Peace*, a 1919 volume critical of the content of the peace treaties that followed World War I, in whose negotiation process he had participated as a member of the British delegation. Although his main activity was on many other fronts—he was already an influential economist as editor of the *Economic Journal*, then the most important economics journal in the world, and a popular polemicist for his interventions in the press—Keynes also continued to work on a volume from the dissertation. Almost in definitive form in 1913, the volume was finally published in 1921 under the title *A Treatise on Probability* (Skidelsky, 1992).

The volume did not go unnoticed by statisticians and mathematicians and was praised as the first systematic treatise on probability in the English language since John Venn's *The Logic of Chance* (1876), which was considered the foundational volume of the frequentist approach that had become dominant in the late nineteenth century. It was from a critique of the frequentist position that Keynes developed his idea that probability is a property of human reasoning, rather than of observed reality. Keynes is still regarded as the first author to systematically study probability as a relation that expresses a degree of belief in a proposition on the basis of evidence, according to an approach to the subject that, following the work of the Austrian philosopher Rudolf Carnap, is usually referred to as the 'logical' approach to probability. The *Treatise* treated probability as a device to deal with situations of less than complete knowledge, and thus explicitly addressed the problem of induction.[10]

We have already observed in the previous chapter that probability analysis is also characterized by the point of view from which it is interpreted. The version discussed in the examples in Chap. 2 is the 'classical' one, in which the probability of an event A is given by the number of cases m in which A can occur out of the n possible cases, i.e., $p(A) = m/n$. Originally developed for the study of games of chance and random mechanisms, this approach was then also applied to all situations in which knowledge of a phenomenon is so limited that there is no reason to assign a different probability to one case than to another. If it is possible to imagine a set of n mutually exclusive cases, which are exhaustive of the possibilities of realization of the phenomenon under scrutiny and all equally possible, then, as Laplace had suggested, there is no sufficient reason not to assign a probability $p = 1/n$ to each case. As a result of this broadening of the scope of analysis—from random mechanisms and games of chance to all situations in which there is no reason to assign a greater probability to one outcome than another—this view became an all-comprehensive theory of probability. This is obviously the a priori probability indicated by Knight.

The 'frequentist' interpretation, conversely, defines probability solely as a function of observable phenomena. Again, it involves a mathematical calculation of probability, as in the classical approach, but without recourse to abstract a priori reasoning. In the frequentist interpretation, probabilistic reasoning is considered to be applicable only to problems where either events are repeated over time or there are a large number of uniform elements that are all observable at the same time. The probability of an event is given by the limit value of the ratio of the number of cases in which the event actually occurs to the total number of observed cases, as the number of cases increases. In the frequentist interpretation, therefore, probability only makes sense if it is based on a specific empirical content, in line with the empiricist approach typical of nineteenth-century English science. This approach was developed in contrast to the rationalism of continental Europe, on which the classical approach was based. There was therefore an explicit rejection of Laplace's idea that one could still derive probabilities under conditions of ignorance. From the frequentist viewpoint, Laplace's procedure would be nothing more than circular reasoning, whereby if all imagined cases are equally possible then they are equiprobable, and therefore incapable of giving a true foundation to the notion of probability.

In criticizing the frequentist position, Keynes ([1921] 1973, p. 93) acknowledged that, for frequentists, "mere ignorance" cannot be the basis for inferences of any kind, and quoted from Venn's *Logic of Chance* the following consideration: "Experience is our sole guide. If we want to discover what is in reality a series of things, not a series

[10] The Humean tradition posits that, in circumstances where demonstrative inferences are rendered unfeasible due to incomplete knowledge, conclusions attain validity solely to a degree of probability. Conclusions derived from such reasoning bear a definite degree of relation to the initial premises, albeit with a lower degree of certainty. Consequently, probabilistic reasoning can be regarded as providing a justification for induction as a scientific method (Gillies, 2000).

of our own conception, we must appeal to the things themselves to obtain it, for we cannot find much help elsewhere." However, according to Keynes, the frequentist approach made probabilistic reasoning too much limited in scope. Knowledge of the relative frequencies of an outcome in a series of similar cases is available in some cases, but the conditions under which such knowledge can be obtained empirically are rarely encountered, and there is no reason to limit the possibility of expressing judgments and opinions in probabilistic form to such a circumscribed class of situations. Keynes called these judgments and opinions "degrees of rational belief."

The proposed identification of probability with statistical frequency, Keynes argued, "is a very grave departure from the established use of words, since in ordinary language it is usual to make a probabilistic judgement even when there is no possibility of empirical observation of a series of cases." As Keynes wrote in the introductory chapter of his *Treatise*,

> In the ordinary course of thought and argument we are constantly assuming that knowledge of one statement, while not *proving* the truth of a second, yields nevertheless *some ground* for believing it. We assert that we *ought* on the evidence to prefer such and such a belief. We claim rational grounds for assertions which are not conclusively demonstrated. We allow, in fact, that statements may be unproved, without, for that reason, being unfounded.

He then continued: "It is not straining the use of words to speak of this as the relation of probability" (Keynes, [1921] 1973, pp. 5–6).[11]

This characterization of ordinary reasoning in probabilistic terms was influenced by the study of the foundations of mathematics in the early work of Russell, for whom mathematics is an extension of logic and all aspects of mathematics are reducible or modeled in logical terms. If mathematics is the logic of certainty, Keynes argued, then probability theory can be denoted as the logic of probability. Just as we normally accept that in some situations it can be argued that a conclusion certainly follows from a premise, so too we can accept that in other cases a conclusion follows in part from a premise, or stands in a probabilistic relation to it. In this extension of demonstrative reasoning to uncertain situations—following Russell's idea and the analytic philosophy of the period, that one can always use a logically coherent language free from contradiction—probability is a partial implication, a relation between a given piece of evidence and a conclusion "in virtue of which, if we know the first, we can attach to the latter a some degree of rational belief," where the adjective rational refers to the logical character of the deduction (Keynes [1921] 1973, pp. 6–7).[12]

---

[11] It should be noted that, in Keynes's approach, probability does not apply to events, but to propositions that express a degree of probability that a hypothesis is true, given a some evidence. According to Keynes, it is necessary to speak of the probability of a proposition *a*, rather than of event *A*, given some evidence. Probability applies to the proposition 'dinosaurs disappeared from the Earth because of the fall of a meteorite,' which can be judged to be true (or false) with some degree of probability, but not to the event of disappearance itself. Despite the philosophical importance of the question, we will ignore this distinction in what follows in order to be able to compare Keynes's argument more easily with those of later authors.

[12] As Carnap (1950) made even clearer, this approach to probability belongs to the field of inductive logic. Probability is considered as an objective notion that allows conclusions to be drawn on the

As we shall see in Chap. 5, the fact that degrees of belief can be represented in probabilistic terms is the basis of any interpretation that insists on the subjective (or epistemic, according to Hacking's distinction recalled in the previous chapter) character of probability. But that these degrees of belief are 'rational' implies a further step, which will be criticized later. According to subjectivists such as Ramsey and de Finetti, probabilities are peculiar to the individual who expresses them, and so there are no degrees of belief in the abstract that are more rational than others, but only degrees of belief that are more or less consistent.

However, at the time of the *Treatise* this subjectivist position was not elaborated explicitly, except for some unarticulated reference to the work of Thomas Bayes. In the *Treatise*, therefore, Keynes concentrated on opposing the idea that the scope of probability theory could be defined by frequentists such as Venn, whose approach Keynes regarded as a step backward from earlier analyses of probability. As we mentioned in the previous chapter, for Keynes the way of arguing put forward by *Port Royal's Logic* indicated that, before the advent of frequentism, it was normal to use probabilistic reasoning also for beliefs and opinions, according to that dual character of probability—aleatory and epistemic—that was admitted in the years of the emergence of the notion of probability that we referred to in the previous chapter. But Keynes was also critical of the classical approach to probability, because the principle of insufficient reason requires the identification of elementary cases—a set of exclusive and exhaustive alternatives to which equal probability of realization can be assigned—which exist in only a small number of situations and would thus limit the scope of probability theory beyond necessary. Both approaches, therefore, he regarded as inadequate to represent probability in its deepest sense, since their acceptance would preclude the idea that probability serves as a universal guide in life, whereas he intended that, in following probability, we act "according to reason" (Keynes [1921] 1973, p. 104).

Moreover, and more importantly, for Keynes both approaches proposed a method of calculating probabilities that assigns them a precise numerical value, whereas in the ordinary situations that interested him, degrees of belief are often expressed imprecisely. Keynes provided the example of judges who have ascertained liability for an injury and who do not know how to determine with precision the probability that the injury was caused by the accused, but who do not consider themselves exempt from exercising judgment. He also mentioned the example of those insurers who insure against any risk, and thus in principle appear to do so because they have assigned a precise probability to the event, yet in practice often do so in an arbitrary manner. The fact that a probability is assigned even in some peculiar cases—such as when certain reinsurance rates are offered for the disappearance of a vessel in distant waters—shows nothing more than sometimes "an occasional gamble may be indulged in," and therefore "the practice of underwriters weakens rather than supports

basis of evidence. While the conclusion does not have the status of certainty of a deduction from a priori statements, it still possesses a claim to some degree of certainty, since it is logically derived from evidence.

the contention that all probabilities can be measured and estimated numerically" (Keynes [1921, 1973, p. 25).[13]

If the probability of an event A is an intermediate value between impossibility, identified by $p(A) = 0$, and certainty, identified by $p(A) = 1$, it is typical of ordinary reasoning, Keynes pointed out, to state that one event in the interval between impossibility and uncertainty is more or less probable than another, so that if A'' is more probable than A' then $p(A'') > p(A')$. But this does not mean that a precise number can always be assigned to all events: for Keynes, the kind of comparability that can be ascribed to probabilities cannot be taken for granted, but must be considered a problematic aspect, since there are not only probabilities that can be expressed with a precise number, but also "non-numerical" probabilities. It is true that many phenomena can be analyzed using Laplace's and Venn's numerical probabilities, but a general theory of probability must also seek to explain phenomena to which strictly quantitative reasoning cannot be applied, and therefore use a qualitative approach where appropriate.

This is why Keynes stated that there is more than one way to define probability along the "path" from impossibility to certainty. There is not only the linear path of real numbers from zero to one—where $0 < p(A') < p(A'') < 1$ and events are completely ordered in terms of probability—but there are also non-linear probability paths, understood as probabilistic representations that are not numerically precise. On a non-linear path, only qualitative comparisons of probabilities can be made, and sometimes even qualitative comparability can be difficult to achieve, if we consider probabilities that refer to different paths. The point about probability paths is illustrated by a figure, the only figure that appears in the *Treatise* (reproduced as in the original in Fig. 3.1). Keynes ([1921, 1973, p. 42) was aware that he was entering an area of probabilistic reasoning that was both controversial and difficult to explain, and that he therefore needed an intuitive representation. His aim was twofold: first, to generalize the notion of probability that was conventional at the time, and second, to show the framework for a possible formal representation of this generalized notion.[14]

What Keynes really meant by non-numerical probabilities remains a subject of debate. The mathematician Russell (1922) showed appreciation for the *Treatise*'s approach, not only because Keynes had followed his logical approach, but also because he himself had argued that a mathematically accurate statement need not necessarily be numerically representable. Thus, qualitative reasoning can be mathematically correct just as much as quantitative reasoning. Russell agreed with

---

[13] That underwriters distinguish between "risks which are properly insurable," because their probability can be estimated within "comparatively narrow numerical limits," and other "risks which cannot be dealt with in this way and which cannot form the basis of a regular business of insurance," as Keynes ([1921] 1973, p. 25) put it, is clearly reminiscent of Knight's distinction.

[14] According to McCann (1994, pp. 41–42), the figure makes it clear that, while the traditional view of probability is cardinal and one-dimensional, Keynes's view is ordinal and multidimensional. More technical, but still suggestive, the consideration made by Kyburg (2003, p. 142) that Keynes's claim was that "the manifold of probabilities is richer than the set of real numbers between 0 and 1."

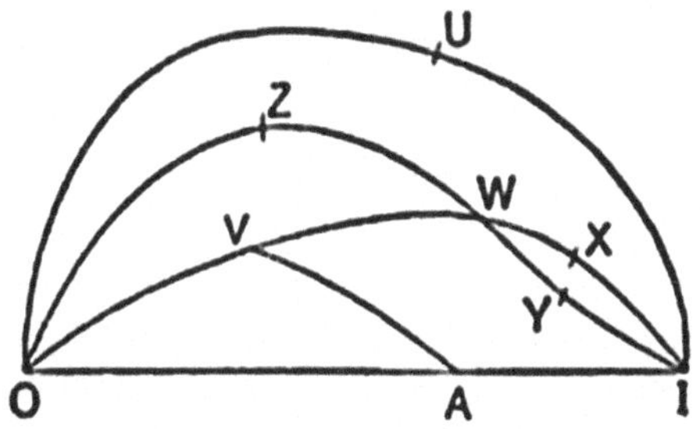

Fig. 3.1 Keynes's probability paths (reproduced from Keynes, [1921] 1973, p. 42)

Keynes's ([1921] 1973, pp. 176–177) contention that in many cases where a numerical measurement is not possible, comparisons of greater or lesser magnitude are possible, thereby rendering numerous other comparisons possible. Although not entirely convinced that Keynes had given a foundation to his notion of probability—a subject we shall deal with later because it was more directly addressed by Ramsey—Russell (1922, p. 121) emphasized Keynes's ability to provide a mathematical structure for his argument, finding "surprising how successful he is in raising his mathematical superstructure upon foundations which might have been thought inadequate to support it."[15]

However, one of the most influential economists of the period, Francis Ysidro Edgeworth—to whom we owe the first representation of an indifference curve, but who was also famous as an advocate of the frequentist approach—did not seem particularly impressed by this aspect of Keynes's account. Edgeworth, who had been editor of the *Economic Journal* before Keynes, argued that it was precisely the observation of similar phenomena in later times, that ever-increasing "statistical experience" emphasized by the frequentist approach that Keynes had rejected, that would refine those probabilities that Keynes had considered non-numerical. The fact that Keynes's probabilities could be non-numerical was thus, if anything, a demonstration that his attempt to abandon the frequentist approach had not been fruitful (Edgeworth, 1922, pp. 279–280).

Before proceeding, a general aspect of the philosophy of probability should be noted. The importance Keynes attached to non-numerical probabilities can be derived from his insistence on ordinary reasoning and the fact that the, in ordinary reasoning, opinions are properties of the subject rather than of the object. Objectivity, for the frequentists, resides in the pure observation of empirical cases, and, for advocates of classical probability, in the properties of the random mechanism under scrutiny. But for Keynes, as for all those who favor an epistemic interpretation of probability—probably Pascal, according to Hacking's reconstruction, but certainly, as we shall see, Ramsey and de Finetti—it is necessary to interpret probability as a property

[15] Keynes devoted an entire section of the *Treatise* to the "fundamental theorems" of his theory, where he endeavored to establish a formal mathematical representation of non-numerical probabilities. This mathematical superstructure, as Russell called it, has been largely neglected in the reading of economists, even those who shared Keynes's view of uncertainty. This same representation, however, would be favorably re-examined in later elaborations, as we shall see in Chap. 6.

belonging to the individual subjects rather than to the objects of their concern. Moreover, considering that at least since the 1930s in economics it has been assumed that the individuals who make decisions do so in relation to their preferences about the objects of choice, in a subjective setting probabilities are regarded as having the same status as preferences. Just as economists have questioned the 'measurability' of the utility with which preferences are represented, a topic that will be addressed in Chap. 4, when faced with choices with uncertain consequences, economists have been compelled to question the 'measurability' of probability, a topic which is reserved for Chap. 6.

Keynes never explicitly returned to the question of the measurability of probabilities, but according to some later interpreters his non-numerical probabilities are in fact nothing more than a further demonstration of his genius (Kyburg, 2003). With his approach, Keynes would have anticipated those scholars who later considered deepening, also from a mathematical point of view, the idea that subjective probabilities are typically intervals of probabilities, that is, not a precise number but an interval between a maximum and a minimum probability value (Brady & Arthmar, 2012). Consider a person who is asked to indicate the probability of a phenomenon such as a collapse of the financial markets—which we have seen in the classification at the beginning of this chapter to be among the most difficult instances to understand in probabilistic terms—and who considers it very unlikely. It seems more realistic to imagine that she would not be willing to indicate a precise probability—e.g., $p(collapse) = 0.01$—but a range corresponding to the indication of a 'very low' probability—so for example $0.005 < p(collapse) < 0.015$. On the one hand, there is the necessity to specify in some way the degree of belief of the person asked to assess the possible occurrence of the event; on the other hand, there is the impossibility of going beyond what is still a qualitative valuation. Hence the appropriateness of the interval, understood as a non-numerical probability.[16]

We will return to these aspects in the chapters on modern decision theory, but it is important to point out here that, although Keynes was not explicitly talking about uncertainty in economics in his *Treatise*, his non-numerical probabilities seem to offer a possible formal representation of Knight's idea that probability cannot

---

[16] The fact that Keynes's reasoning is also mathematically sophisticated, is supported by the observation that when probabilities are numerical they express a complete ordering—it is always possible to know, for each pair of events A' and A'', whether $p(A'') > p(A')$ or $p(A'') < p(A')$ or $p(A'') = p(A')$. But if the probabilities are non-numerical, Keynes observed, the ordering is only partial, because it may happen that for some pairs of events it cannot be determined whether one is more or less probable than the other, or whether they have the same probability. In fact, if non-numerical probabilities are interpreted as intervals of probabilities they may even overlap, in whole or in part, making probabilistic reasoning much more complex. For example, if it holds for a given pair of events A' and A'' that $0.2 < p(A') < 0.4$ and $0.4 < p(A'') < 0.5$, then A'' is more probable than A'. But if it is true that $0.2 < p(A') < 0.4$ and $0.3 < p(A'') < 0.5$, then no order between the probabilities can be established. This is a possible account of what distinguishes quantitative from qualitative probabilistic reasoning, with Keynes discussing the qualitative (Runde, 2001). The non-comparability emphasized by Keynes may therefore depend on the intersection between the intervals, not necessarily on the lack of ordering in itself, as in the famous example of forecasting the weather when the clouds are black but the barometer is high (Keynes, [1921] 1973, p. 32).

represent "true" uncertainty. Since, according to Keynes, numerical probabilities do not always exist when it comes to degrees of belief, there seems to be a clear assonance between Keynes and Knight on this ground. However, whereas Knight left the field of investigation concerning what he called estimates essentially undefined, Keynes delved into it through a specific probabilistic analysis, examining the possibility that a calculus of non-numerical probabilities could be elaborated. We shall see in Chap. 5 that the extent to which subjective probabilities could be considered definite—that is, numerically precise—was a matter of crucial disagreement between Keynes and his more critical reviewer, the young Cambridge mathematician Ramsey.

Moreover, the closeness of the two analyses is also demonstrated by a related aspect, that concerning the confidence with which a given degree of belief is formulated. We saw in the previous section that Knight rejected the idea that his estimates could be probabilities, however subjective, and that he did so also on the basis of the different confidence that can typically be attached to estimates as opposed to probabilities. In practice, Knight argued, objective probabilities are precisely identifiable, whereas subjective probabilities are not; subjective probabilities, if they exist, would imply the necessity to assess the degree of confidence with which they are formulated, which by definition is less than the maximum degree of confidence that can be attributed to objective probabilities. Again, after introducing the issue, Knight did not elaborate a proper probabilistic analysis; Keynes, on the contrary, devoted an entire chapter of the *Treatise* to what he called the "weight of arguments."

After explaining what he meant by probability as a representation of a "rational" degree of belief, Keynes emphasized that the evidence on which some probability relations are based may differ, and thus influence the way an individual views a given probability. Keynes ([1921] 1973, p. 82) gave a very simple but suggestive example from the analysis of games of chance, which would later be taken up in the debates on modern decision theory. Suppose we have before us an urn containing a given number of balls, which may be either white or black, but we do not know in what proportion. Since we can assume that each ball is either white or black and that the events (of a ball being either white or black) are mutually exclusive, we can apply the principle of insufficient reason to say that the probability of drawing a white ball from this urn is $p(white) = 1/2$. Now suppose we have before us another urn, containing the same number of balls as the previous one, either white or black, but of which we know that the proportion of balls is 50% white and 50% black. Even in this second case $p(white) = 1/2$. However, Keynes argued, it seems clear that there is "a more substantial basis upon which to rest our conclusion" about this second probability valuation. The probability is the same in both cases, but in the second case a greater "weight of argument" supports the conclusion that the probability is 1/2, or, to use Knight's term, the confidence in the probability valuation is greater in the second case.

Keynes specified that he was referring to two different properties of probabilistic reasoning: first, there is probability, which by definition is based on evidence more or less favorable to the occurrence of the event; second, there is weight, which also depends on evidence, but in the sense of the amount of evidence available, possibly measured by "the degree of completeness of the information upon which a probability is based" (Keynes, [1921] 1973, p. 345). Obtaining new evidence may lead us to change the probability in the sense of making its value higher or lower, but in any case it makes us more confident in identifying the probability value.

There are at least two reasons why Keynes's insistence on the weight of the argument must be noted. First, there is an interpretative question concerning the continuity that can or cannot be envisaged between the young Keynes's probability theory and his later contributions to economic theory. We must remember that we are examining a dissertation from 1908, possibly in its final form for printing just before World War I, even though it was published in 1921, whereas the *General Theory* was published in 1936. It is precisely in connection with his notion of weight that Keynes made a precise reference to the *Treatise* in the *General Theory*, with the chapter on the weight of argument indicated as the place where the meaning of the uncertain bases for formulating expectations that he intended to emphasize in the *General Theory* were properly explained. As we shall see in the next section, a central aspect of the reasoning about uncertainty in economics that we find in the *General Theory* is the confidence in the assumptions about the performance of financial markets on the basis of which investment decisions are made.

The second reason relates to a possible interpretation of the role of the weight of argument suggested by Keynes himself, when, in the *Treatise*, he considered how probability theory could be used in decision-making. In this context, Keynes suggested that if the weight of the argument is maximum, then the numerical probabilities of the traditional calculus can be used to make decisions; but if the weight of the argument is lower, then the probabilities are non-numerical, and the question of how to make decisions obviously becomes more complicated. Keynes ([1921] 1973, pp. 348–349) went so far as to illustrate his reasoning with a decision coefficient, analogous to probability, but dependent on both probability and the weight of argument, and therefore not satisfying the additive property of a probability measure.[17] But he then concluded that he doubted that the solution could be found in some complicated function of probability.

We have observed that Keynes's aim in the *Treatise* was to make a contribution to the theory of probability in itself, proposing a new approach intended to overcome the frequentist idea. For example, the word uncertainty never appears in the *Treatise*

[17] In place of a calculation of the expected value based on probability alone, he suggested a calculation based on a probability modified according to the weight, which he wrote as $c = p[2w/(1 + w)]$ (reproduced here in a slightly modified version to avoid discussing the peculiar way in which Keynes took aversion to risk into account). Simplifying, he expressed the case in which the degree of completeness of the information is believed to be maximum with $w = 1$, and the case in which this is not true with $0 < w < 1$. If the weight is maximum, the probabilities are reliable weights, so that, if $w = 1$ then $c = p$, and the decision coefficient is nothing but the ordinary probability weight, as is usual in calculating expected utility. But if the degree of completeness of the information is lower, then $c < p$. For example, if $w = 3/4$ then $c = (6/7)p$. The coefficient c can be understood, in modern terms, as a 'decision weight,' that is, a distortion of the probability p assigned to the occurrence of a given event, depending on the degree of completeness of the information. It has been noted that this idea of a numerical probability distortion anticipates one of the assumptions of Kahneman and Tversky's 'prospect theory,' to which we will return briefly in Chap. 7 (Brady, 1993).

to refer to uncertainty in economics, in the style of Knight. But it is also true that his insistence on the importance to use probability theory to represent (rational) degrees of belief was due to the idea that probability is a "guide of life," a tool for making decisions. This approach to probability theory that emphasizes its role in decision-making is so important that Keynes devoted an entire chapter to it, entitled "The Application of Probability to Conduct." Although in a sense peripheral to the main purpose of the volume, this chapter is obviously of great interest from an economic point of view, because it contains the first detailed Keynesian description of how to make decisions under conditions of uncertainty, and thus allows us to outline the features of a peculiar decision theory derived from his thoughts on probability.

Although in the years in which he wrote the *Treatise* Keynes had to deal with the ethics of Moore—the most philosophically relevant figure within the circle of the *Apostles*—in order to deal with the problem of choice, Keynes took as his point of departure utilitarian ethics, which he indicated as the "usual ethical theory" of the time. Faced with the ethical question of what is right to choose in a given situation, utilitarian ethics proposes as the criterion to be followed the maximization of the mathematical expectation of the outcomes of actions, thus accepting a more or less explicit idea of the maximization of expected utility.

This ethical position was opposed by Moore, for whom the probabilities of actions whose outcomes occur in the distant future could not be taken for granted, and therefore the associated consequences should not be taken into account in the act of choosing what is ethically better to do. According to Keynes, on the other hand, even if the future results of a given action are very uncertain, they must always be taken into account, because "we have a genuine probability, even when the evidence upon which it is founded is slight" (Keynes, [1921] 1973, p. 342). Moore was therefore relying on a flawed notion of probability, the frequentist notion; but since probability is a matter of subjective beliefs, not objective reality, both his critique of utilitarian ethics and his idea of what should be meant by ethical conduct had to be revised.[18]

However, the fact that the multiplication of probabilities by the utilities of the consequences of an action was generally a more convincing criterion for decision than

[18] In Moore's approach, while it is true that action for the attainment of the moral good must take into account the probability with which a particular action can achieve it, if the effects of an action are linked to a future realization the probability of which cannot be known, this should be disregarded. Neglecting the future positive effects of an action, in favor of a present good, could mean that a higher utility (goodness) that can only be achieved in the distant future cannot be preferred to a lower utility achievable in the immediate future, simply because the probability of the latter is known while that of the former is not. In arguing that the utilitarian approach could not be applied, Moore therefore proposed to follow conventional rules of morality, rather than those dictated by logic. Keynes opposed this view, which would exclude any behavior that deviates from existing moral conventions, such as the pursuit of goals with immediate consequences that current morality does not contemplate, but that may later prove to be well justified (Raffaelli, 2006). Among the young members of the Apostles, and then again in the Bloomsbury circle in later years, the critique of current morality was considered to be a real mark of the intellectual's ability to confront reality. According to some interpreters, this elitist component of Keynesian thought serves to explain his approach when, as an economist, he sought to counter the application of almost unanimously accepted economic policies in the 1920s and 1930s (Skidelsky, 1992).

Moore's, did not make it any less controversial. To this end, in the chapter on conduct, Keynes reminded the reader of the two main characteristics of his probabilities: first, probabilities always exist, but they are not necessarily precise numbers; second, their actual value must be weighed against the weight of the argument, depending on the degree of completeness of the available evidence. With Keynesian probabilities, one cannot use "a mere process of arithmetic to determine which of the alternatives ought to be chosen."

As already noted, the idea of entrusting the solution of a choice problem to a "precise numerical method" was difficult for Keynes to pursue. Hence his conclusion, reinforced, as we shall see in the next section, by a number of passages from the *General Theory*, that a strictly mathematical approach to choice theory—an approach that privileges numerical precision over the explanatory value associated with a non-numerical, logical approach—is inadequate to provide the answers that are required in some contexts, such as the economic one.

This does not mean, however, that one should rely on a Moore-like moral approach and abandon the idea of choosing as rationally as possible. Relying on conventions may sometimes be necessary, but the importance of probability is reaffirmed. After a series of precise critiques of the conventional use of probability, Keynes concluded that the importance of probability lies in the fact that "it is rational to be guided by it in action" and that "in action we *ought* to take some account of it" (Keynes, [1921, 1973, p. 356). Human action cannot always rely on precise numerical calculations, but this does not mean acting without a criterion, because (probabilistic) logic underlies rational thought, and even in a (probabilistically) complex context, human action must be conscious, not extemporaneous. While it is true that "mathematical reasoning now appears as an aid in its symbolic rather than in its numerical character" (Keynes, [1921] 1973, p. 349), a correct probability theory nevertheless provides the basis for making economic decisions, even under conditions of uncertainty.

3.4 Uncertainty in Keynes

We have already observed that probability and uncertainty are closely related concepts. This is mainly because, for any uncertain phenomenon, we are interested in knowing the value of the probability with which it can occur, however controversial that value may be. But the preceding lengthy disquisition on probability in Keynes might appear misplaced. In fact, Keynes's scholarly activity at the time of the publication of the *Treatise* was apparently focused on quite different issues. Keynes was intent to counter what in the economic theory of the time was at odds with the need, which he saw as pressing in the post-war period, to adopt economic policy decisions that would not hinder Britain's growth. Both in his writings on monetary theory and in polemical articles in the press, he argued against the country's return to the gold standard at pre-war exchange rates. He believed that the external constraint imposed by their adoption would severely limit the ability to stabilize domestic prices, with deleterious effects on employment. This he argued despite still accepting the standard view of interest rate determination in those years, an acceptance that made his way of arguing essentially internal to what he would later criticize as a "classical" approach to economics in his *General Theory* (Keynes, [1936] 1973, p. 1).

It was in the early 1930s that Keynes's representation of the workings of the economic system evolved into his well-known revolutionary suggestions: faced with the inability of the standard model to suggest effective measures to get out of the Great Depression, Keynes formulated a macroeconomic model that would allow economic policy, particularly fiscal policy, to be integrated into analytical reasoning. His criticism of the "classical" economists, and their belief in the self-regulating capacity of the economic system, was expressed in particular in an opposition to the so-called Say's Law, according to which any problem of demand deficits could only be sectoral and not of the system as a whole. To this criticism Keynes then linked the contention that the interest rate is determined in the money market, and not in the market for loanable funds as the classicals believed. Both aspects turned out to be related to Keynes's view of uncertainty and this allows us to isolate its role within the *General Theory*. The aim of this section is to highlight the common thread that connects the probability theory of the early years with the economic theory of the mature period, that which is still taught in macroeconomic textbooks today.[19]

The Keynesian analysis of uncertainty is first developed in Chap. 12 of the *General Theory*, entitled "The State of Long-Term Expectation." Having examined in the previous chapter the real components that drive investment activity—with the planned level of investment in a given period being determined by comparing what he called the "marginal efficiency of capital" with the market rate of interest—Keynes turned to the question of why in certain periods the level of investment does not correspond to the planned level. The marginal efficiency of capital is derived from the calculation of entrepreneurs, who determine the internal rate of return for a new investment that they intend to undertake as a function of the expected future returns relative to present costs. This calculation is in principle a rational one, analogous to what, Keynes ([1936] 1973, pp. 142–143) reminded the reader, one can find in Marshall's notion of the "marginal utility of capital" or in Irving Fisher's notion of the "rate of return over cost." But future returns are based on expectations, and it is therefore to these expectations, and the confidence with which they are formulated, that one must look in order to explain the performance of private investment. This is the aspect that Keynes felt to be substantially neglected in the economic analyses of the time, which were developed as if all future variations could be correctly anticipated. As he would summarize after the publication of the *General Theory*, with reference to the classicals, including his contemporaries Edgeworth and Arthur Cecil

---

[19] The assumption of a continuity of thought between two such different works, written almost thirty years apart, is an interpretative operation that is not readily apparent. Since its formulation in Carabelli (1988) and O'Donnell (1989), it has attracted opposite reactions, as documented in the collected volume edited by Runde and Mizuhara (2003). The alternative interpretative hypothesis, that of a clear discontinuity, is associated with Keynes's reaction to Ramsey's strongly critical comments on the *Treatise*, but we defer its assessment to Chap. 5, which deals with the impact of Ramsey's subjectivism on decision theory.

Pigou, "the calculus of probability, though mention of it was kept in the background, was supposed to be capable of reducing uncertainty to the same calculable status as that of certainty itself" (Keynes, 1937, p. 213).[20]

According to Keynes, it is crucial to acknowledge that with increasingly organized stock and financial markets, the present value of investments is essentially (re)valued day by day. It is true that transactions in markets facilitate the financing of investments, but the activity of buying and selling that values an existing investment also has an indirect effect on the valuation of new investments. An investment should be valued by those who operate on the markets according to traditional canons, linked to real assessments of the intrinsic value of the asset, the so-called fundamental values, that is, according to the "genuine expectations of the professional entrepreneur." This is already a daunting task in itself, because it presupposes estimates of factors that typically manifest themselves in future periods. But to this must be added that investments have to be evaluated on the basis of expectations that are generated by the daily fluctuations of the stock market, which depend on the choices of a whole range of different financial actors, often driven by different objectives. Indeed, Keynes observed, those who operate in the markets are well aware that they have to take into account various elements beyond those that determine fundamental values. But economists do not normally take this into account in their analyses.

Chapter 12 of the *General Theory*, with its description of the role of expectations in the stock market, is one of the most quoted pieces of Keynes's entire oeuvre. His description of the discrepancy between what financial markets are supposed to serve and what actually happens in those markets is so accurate and vivid that it is still considered by many to be unsurpassed. Keynes examined the behavior of the different types of investors operating in the market. First and foremost, there are the activities of those who wish to invest in the real business, to which he assigned the term "enterprise," for the activity of "forecasting the prospective yield of assets over their whole life." This is a complex activity due to the "the extreme precariousness of the basis of knowledge on which our estimates of prospective yield have to be made." In fact, the knowledge of the factors which will govern the return on an investment after years is "usually very slight and often negligible," as in the case of the ten-year return "of a railway, a copper mine, a textile factory, the goodwill of a patent medicine." This means that "those who seriously attempt to make any such estimate are often so much in the minority that their behaviour does not govern the market" (Keynes, [1936] 1973, pp. 149–150).

[20] Keynes distinguished between short-term expectations, for the choices of entrepreneurs when the capital endowment is given, and long-term expectations, for the investment in new capital. In the short run, it is possible to represent each choice by means of a "certainty equivalent," that is, by calculating the certain value that corresponds to the expected profit based on the probabilities with which entrepreneurs expect to obtain returns from their activity. Again, Keynes hinted at possible problems with the reliability of the entrepreneur's expectations, but did not dwell on them. Long-term expectations, on the other hand, require specific analysis, precisely because of the difficulty of identifying certainty equivalents. In Keynes's view the classical economists used short-term methods to form expectations that were then used in their long-run models, a methodology resurrected in the neoclassical macroeconomics of the 1970s (Hoover, 1997).

Moreover, the valuation of an investment cannot be based on a "strict mathematical expectation," as expectations depend not only on the probability of the realization of expected future returns, but also on the confidence with which they are formulated and "it would be foolish, in forming our expectations, to attach great weight to matters which are very uncertain." The reference to the *Treatise* that we mentioned in the previous section was made in a footnote to this sentence, where Keynes ([1936] 1973, p. 148) clarified: "by 'very uncertain' I do not mean the same thing as 'very improbable'." Thus, it is not only a question of what probability value is to be attached to a future event, but also of the reliability of that probability, something that Keynes was evidently still thinking as representable in terms of the weight of argument, a representation that, in principle, is amenable to mathematical terms (Basili & Zappia, 2021).

But, secondly, there is the speculative activity carried out by those traders who buy in the market not because they have a precise idea of what the fundamental values of the asset are, but for short-term reasons. "Speculation," as opposed to "enterprise," is based on the fact that some traders believe that other traders will also buy that particular asset. For these speculators, the important thing is to correctly anticipate the market trend, not to identify which investment really has the capacity to generate profits in future. And if these traders are in the majority at a particular point in time, the market will move in the direction they think it should move, without any genuine connection to the underlying real structure of the economy. Obviously, market values calculated in this way—expressing what Keynes called a "conventional method of calculation" rather than an attempt to calculate an intrinsic value—have a precarious basis and are therefore subject to sudden and often erratic changes, which may affect the entire economic system.[21]

The uncertainty surrounding business activity is thus both an uncertainty inherent to the individual investment and one linked to a market environment that changes according to what speculation dictates. However, the analysis of such an environment, so complicated to rationalize from a theoretical point of view, is not developed by classical economics, which, on the contrary, simply ignores it: faced with what Keynes considered to be an obvious difficulty of calculation, economists confined themselves to the standard calculation, as if the difficulties did not exist. According to Keynes, the classical economists did nothing more than follow the less shrewd operators in the market: unable to assess what a future as complex as that generated by this way of investing in the stock market would be like, they relied on the idea

[21] The example Keynes used to clarify what he meant by a "convention" is famous: in the markets, traders often behave as if they were taking part in a game in which the winner is the one who guesses which photograph will win a beauty contest, among a series of photographs from which the most photogenic person must be selected. Players do not have to think about which photograph they think is the best, but which photograph the majority of the players will think is the best, with all players looking at the problem from the same point of view, potentially creating a situation of infinite regress (Keynes, [1936] 1973, p. 156).

that in the immediate future things would go on as if nothing had changed, that is, to simplify, as if the probabilities used today would still be appropriate tomorrow.[22]

The analysis in Chap. 12, however compelling, was limited to financial markets. But for Keynes the issue was evidently more general, and concerned the tendency to consider uncertainty as simply as risk in the study of all economic problems, which for him was typical of the traditional approach in economic theory. This is the message entrusted to the summary of the *General Theory* that Keynes proposed in the 1937 article that we quoted in the introductory chapter of this volume. In summarizing the sense in which uncertainty conditions economic activity, Keynes (1937, pp. 212–213) insisted that, about matters of which "there is no scientific basis on which to form any calculable probability whatever," one cannot proceed by pretending that the question is one of risk. Battling against this approach, defined as a "false rationalization [which] follows the lines of Benthamite calculus"—a reference to the father of utilitarian ethics, Jeremy Bentham, which brings us back to the themes of the *Treatise*—Keynes placed uncertainty as the first of the "simple fundamental ideas" underlying his theory.

As it emerges from the list of cases made in the quotation from the 1937 article with which we opened the volume, Keynes had in mind a gradation of decision-making contexts, ranging from situations of risk to situations of extreme uncertainty.[23] They range from risk, such as when one can assume equal probability of elementary outcomes—e.g., playing roulette or buying a bond whose early redemption rules are known—to moderate uncertainty, in which one can partly rely on the observation of frequencies—e.g., life expectancy or weather forecasts. But then there are the situations that Keynes was most interested in, the situations of true uncertainty, those in which the same reasoning as above cannot be followed, due to the lack of scientific bases that allow one to conceive of a "calculable probability"—e.g., "the prospect of a European war [...] or the obsolescence of a new invention." The distinction between cases that can be labeled as either risk or uncertainty evidently accords well with Knight's analysis. Hence, as we have already seen, the custom of

---

[22] In this context not even professional investors can make useful choices for the economy, but at most devote themselves to beating the market. Keynes was highly critical of the New York Stock Exchange, which he pointed to as the prototype of this (mal)functioning of the markets: "when he purchases an investment, the American is attaching his hopes, not so much to its prospective yield, as to a favourable change in the conventional basis of valuation ... he is, in the above sense, a speculator. Speculators may do no harm as bubbles on a steady stream of enterprise. But the position is serious when enterprise becomes the bubble on a whirl-pool of speculation. When the capital development of a country becomes a by-product of the activities of a casino, the job is likely to be ill-done" (Keynes, [1936] 1973, p. 159).

[23] We reproduce here Keynes's description of the cases (1937, pp. 213–214): "By 'uncertain' knowledge, let me explain, I do not mean merely to distinguish what is known for certain from what is only probable. The game of roulette is not subject, in this sense, to uncertainty; nor is the prospect of a Victory bond being drawn. Or, again, the expectation of life is only slightly uncertain. Even the weather is only moderately uncertain. The sense in which I am using the term is that in which the prospect of a European war is uncertain, or the price of copper and the rate of interest twenty years hence, or the obsolescence of a new invention, or the position of private wealth-owners in the social system in 1970. About these matters there is no scientific basis on which to form any calculable probability whatever. We simply do not know."

considering Knight and Keynes as the founders of the distinction between risk and uncertainty.

However, Keynes went further than Knight. On the one hand, he reiterated that to assess the influence of extremely uncertain phenomena one cannot use "pretty, polite techniques," as classical economists usually did. Here he cited a tradition beginning with David Ricardo and Marshall and that he saw still alive in Edgeworth and Pigou: as already noted, even when they admitted that expectations about the future could be disregarded, they did so by assuming that expectations were given in a calculable form, as if the calculus of probabilities were capable of "reducing uncertainty to the same calculable status as that of certainty." On the other hand, Keynes stated that "tho[ugh] this is how we behave in the market place, the theory we devise in the study of how we behave in the market place should not itself submit to market-place idols" (Keynes, 1937, p. 215). Thus, Keynes's suggestion was that we need to imagine alternative computational techniques for making decisions. In this sense, emphasizing the possible continuity between the *Treatise* and the *General Theory*, one can imagine that when Keynes said that some phenomena are particularly uncertain because of the impossibility of conceiving "any calculable probability," he meant by "calculable" a numerical probability.

This would leave the discourse open to the possibility of a representation of uncertainty through "non-numerical" probabilities, as it seemed to be confirmed by an important correspondence in the period following the publication of the *General Theory*, which we briefly mentioned in the introductory chapter. To Hugh Townshend, who was working on a review for the *Economic Journal* of a book on expectations and investment, Keynes ([1937–1938] 1979, pp. 288–289) wrote that the question of which criterion of choice was apt to deal with uncertainty was "a very important and interesting one, often in my mind." Keynes suggested that he found it correct to refer somehow, as Townshend had done, to the notion of certainty equivalent. But he then concluded:

> a main point to which I would call your attention is that, on my theory of probability, the probabilities themselves, quite apart from their weight or value, are not numerical. So that, even apart from this particular point of weight, the substitution of a numerical measure needs discussion.

Since this is the only exchange in which the subject of decision criteria under uncertainty is taken up by Keynes after the *General Theory*, it is difficult to underestimate its significance (Zappia, 2015).

Indeed, despite a description of financial markets that highlighted their dubious rationality, Keynes had already insisted in Chap. 12 on the necessity to find alternative solutions. Commenting on the issue of speculation, Keynes ([1936] 1973, p. 155) made it clear that "the social object of skilled investment should be to defeat the dark forces of time and ignorance which envelop our future." This statement is a prelude to Keynes's wish, formulated at the end of the chapter, that the economic policy authority intervenes with targeted public investment, if private investment is not undertaken because of uncertainty. And on the question of the "human nature" that gives rise to the behavior of the traders he stigmatized, he made it clear that

> we should not conclude from this that everything depends on waves of irrational psychology .... We are merely reminding ourselves that human decisions affecting the future, whether personal or political or economic, cannot depend on strict mathematical expectation, since the basis for making such calculations does not exist. (Keynes, [1936] 1973, pp. 162–163)

In Keynes's perspective, therefore, uncertainty is the main problem to focus on for an adequate analysis of the (mal)functioning of market economies. But the solution to the problem of uncertainty is never to give in to irrationality, although this is something that many economic agents actually end up doing. The irreducibility of uncertainty to risk is never an indication that uncertainty should not be analyzed as rationally as possible. Those who really invest because they follow the fundamentals at times when no one else does, or who feel a "spontaneous urge to action rather than inaction"—according to the famous Keynesian reference to the "animal spirits" of the entrepreneur (Keynes [1936] 1973, p. 161)—are economic agents who also make a calculation, albeit an evidently different one from the conventional one.[24] *A fortiori*, Keynes argued, even policymakers, who are the only actors on whom we can rely in severe crises, cannot but refuse to ascribe rationality to market outcomes, when it is evident that they are guided by conventional judgments. It is precisely in situations of real uncertainty that policymakers' action is essential to subvert behavior driven by market idols; such an action must be based on a long-term vision that favors the "general social advantage" (Keynes [1936] 1973, p. 164).

Before concluding, it is necessary to recall another important connection that Keynes stressed in the *General Theory* between probability and uncertainty, namely the relationship between expectations and the interest rate. Keynes considered this relationship when he introduced the notion of "liquidity preference" to explain the willingness (or unwillingness) of the investors holding money for precautionary/speculative purposes to help finance private investment. As is well known from macroeconomics textbooks, for Keynes the demand for money for precautionary/speculative purposes is inversely related to the interest rate because it incorporates investors' expectations of possible future fluctuations in the interest rate. A 'low' interest rate relative to a supposedly 'normal' interest rate is associated with a 'high' value of assets in the secondary market and generates the expectation of a possible future loss if assets are held rather than money, and vice versa.

However, what the 'normal' interest rate is—what equilibrium interest rate is expected to prevail in the money market—depends on the expectations about the future profitability of investments, which we have seen Keynes called the "state of

---

[24] That of "animal spirits" is a rhetorical figure often recalled in the literature on Keynes, mostly to place emphasis on the irrational attitude of many types of individuals operating in markets (Akerlof & Shiller, 2009). However, to argue that the animal spirits of entrepreneurs signal an irrational attitude in a broad sense does not seem to be Keynes's main message (Dow & Dow, 2011). If investors are economic agents who assign a range of probabilities to a given event that may condition the returns on their investment, then the animal spirits can be interpreted as the determination to assign to a positive (negative) event the maximum (minimum) probability among those contemplated, in line with the *Treatise*'s idea of representing imprecise probabilities by means of an interval. In what sense Keynes might also have had in mind a 'rational' criterion for dealing with this aspect of decision-making is analyzed in Zappia (2016).

long-run expectation." In a very uncertain situation, the premium required to get rid of money and invest in assets is higher than that justifying a given 'normal' interest rate: the money demand curve shifts upwards due to uncertainty and, for a given money supply, the same 'normal' equilibrium interest rate is now higher than before. According to Keynes (1937, p. 216),

> partly on reasonable and partly on instinctive grounds, our desire to hold Money as a store of wealth is a barometer of the degree of our distrust of our own calculations and conventions concerning the future … . The possession of actual money lulls our disquietude; and the premium which we require to make us part with money is the measure of the degree of our disquietude.[25]

In situations of 'Keynesian' uncertainty, as we have illustrated, not only is it difficult to formulate accurate probabilities, but also their degree of reliability is low. This prompted Keynes to focus on non-numerical probabilities. Consequently, in order to induce holders of money to invest in investment-supporting financial assets when they are unable to form accurate forecasts about the future, it is not enough to merely compensate for the usual risk inherent in the investment; an additional compensation linked to the state of uncertainty must be offered. According to Keynes ([1936] 1973, pp. 239–240) liquidity is a demand for flexibility in the possibility of varying one's portfolio between assets associated with long-term investments (in new capital), on which high returns can be anticipated, and low-yielding assets associated with the debt that companies and governments make for normal operation in the short term (at constant capital)—essentially between shares and bonds. The more uncertain the situation is, the less attractive it will be for who holds money to invest it in financial instruments aimed at financing real investments in new capital, instead of keeping to hold securities or money even though they guarantee a very low interest or no interest at all.

Returning to the money market, given a certain money supply in the system, the equilibrium interest rate that is determined with respect to the precautionary/speculative demand for money will be higher (lower) depending on the higher (lower) remuneration that is demanded for liquidity at a given time. If the target of monetary policy is a low interest rate, in order to induce a high level of real private investment, a situation in which the demand curve for speculative money shifts upwards due to the effects on the increased demand for liquidity may render any increase in the money supply ineffective. This is what typically happens in situations of permanent depression of the economic system and thus explains the inability of the system to recover after a depression, up to the extreme situation of a liquidity trap (Hayes, 2010, pp. 69–70).

---

[25] This aspect is reiterated in the above-mentioned correspondence with Townshend: "I am quite tempted to associate the risk premium with probability in the strict sense, and the liquidity premium with what in my *Treatise on Probability* I called 'weight' […] A liquidity premium is a payment, not for the expectation of a higher tangible income at the end of the period, but for a heightened sense of comfort and confidence during the period" (Keynes, [1937–1938] 1979, pp. 293–294). The importance of the *Treatise* for an adequate understanding of the role of liquidity in the *General Theory* is assessed in Runde (1994)) and Vercelli (2018).

We do not have space here to go into more detail on uncertainty in the money market as distinct from the financial market. But Keynes's understanding was that the effect of uncertainty in terms of preference for liquidity is compounded by the effect of uncertainty on the valuation of future returns on an investment, since a more pessimistic view of the latter can only induce more money holdings. In situations of extreme uncertainty, the "fetish of liquidity"—another often-remembered Keynesian figure, indicating the attitude of investors to holding "liquid" assets (Keynes, [1936] 1973, p. 155)—conditions both the financial and money markets, making the financing of private investment activity difficult if not impossible.

3.5 Concluding Remarks

In this chapter, we have examined the thinking of the two authors who have contributed more than anyone else to shape awareness of the need to theorize about uncertainty in economics, and thus to make it central to the analysis rather than to relegate it to marginal considerations. We have done this by dwelling on the consonant aspects of their independent analyses. The distinction between risk and uncertainty as conceived by Knight is clear-cut, and can be summarized in the idea of irreducibility, that is, the idea that uncertainty cannot be reduced to risk because, unlike risk, it cannot be measured. Keynes had been moving in a very similar area since the *Treatise*, but he did so by approaching the question from the point of view of probability theory, distinguishing between situations in which probabilities are numerically precise and those in which they are not. Knight argued that beyond objective probabilities one could only formulate estimates, degrees of belief that are undefined probabilistic judgments. Keynes proposed a specific probabilistic analysis of degrees of belief, an analysis that was innovative at the time because it opened up the discussion about the sense in which probabilities could be interpreted as subjective.

In a sense, as we have tried to illustrate, Keynes distinguished between a quantitative approach which he was critical of, and a qualitative approach to which he nevertheless endeavored to give analytical content. Since the distinction can also be understood as a question of the confidence with which a definite rational degree of belief can be formulated, the former case can be said to correspond basically to Knight's risk, while the latter corresponds to Knight's uncertainty. We will see in the following chapters that both Knight's distinction and Keynes's analysis would be successfully criticized later, particularly from the 1950s onward. But we will also see that they have remained alive in some later interpretations and eventually re-evaluated in contemporary decision theory.

We have also observed that the Keynes of the *General Theory* explicitly addressed the issue of uncertainty in economics. Keynes basically agreed with Knight on the difference between an economic theory that assumes some form of omniscience on the part of agents and one that admits that this way of theorizing basically means ignoring the differences between risk and uncertainty. But even in this second case,

Keynes's analysis is more thorough, since he admitted that there are different degrees of uncertainty, not only the immeasurable uncertainty bordering on total ignorance. In relation to this complexity, the response of economic agents is not uniquely determined. Keynes tried to outline how the situation of uncertainty induces different agents to behave differently. In financial markets, in particular, some traders rely on conventions, either by ignoring the difficulties of calculating probabilities—thus operating as if they were in a situation of risk rather than uncertainty—or by trying to align themselves with market sentiment. Other traders, though aware that this makes little sense, adapt to the situation, but they also try to anticipate changes in the current convention, the fragility of which they recognize. Still others nevertheless try to assess fundamental values and are willing to take the risk of going against the market, as is typical of the animal spirits of capitalism. In this extremely complex context, the idea is justified that there are moments in economic history, such as the Great Depression of the 1930s, when only the economic policy authority can succeed, because the actual working of markets depress any possible private initiative.

In conclusion, although Keynes and Knight share a general idea that is still identified in economics today with the idea that 'Knightian' uncertainty requires a specific analysis—an idea opposed by the decision theory that would consolidate after World War II—it is to Keynes more than Knight that one must look to when examining heterodox attempts to study the decisions of economic agents under conditions of uncertainty irreducible to risk. This is all the more true if, as in the reconstruction we have proposed, the probability analysis of the *Treatise* is used to give analytical content to the uncertainty of the *General Theory*, since there is a common thread of discourse between the two works, leading from 'non-numerical' probabilities to 'Keynesian' uncertainty.

References

Akerlof, G., & Shiller, R. (2009). *Animal spirits*. Princeton University Press.

Arrow, K. J. (1951). Alternative approaches to the theory of choice in risk-taking situations. *Econometrica, 19*, 404–437.

Basili, M., & Zappia, C. (2021). Financial markets and Keynes's long-term expectations. *Cambridge Journal of Economics, 45*, 1047–1067.

Birner, J. (2015). Complexity and social sciences. *On the Horizon, 23*, 100–106.

Borch, K. (1976). The monster in Loch Ness. *Journal of Risk and Insurance, 43*, 521–525.

Brady, M. E. (1993). J.M. Keynes's theoretical approach to decision-making under condition of risk and uncertainty. *The British Journal for the Philosophy of Science, 44*, 357–376.

Brady, M. E., & Arthmar, R. (2012). Keynes, Boole and the interval approach to probability. *History of Economic Ideas, 20*, 65–84.

Carabelli, A. M. (1988). *On Keynes's method*. Palgrave Macmillan.

Carnap, R. (1950). *Logical foundations of probability*. University of Chicago Press.

Dow, A., & Dow, S. C. (2011). Animal spirits revisited. *Capitalism and Society*, 6, article 1.

Edgeworth, F. Y. (1922). The philosophy of chance. *Mind, 31*, 257–283.

Fiorito, L. (2016). A certain amount of 'recantation': On the origins of Frank H. Knight's antipositivism. *History of Political Economy, 48*, 1–34.

Gilboa, I. (2009). *Theory of decision under uncertainty*. Cambridge University Press.

Gillies, D. (2000). *Philosophical theories of probabilities*. Routledge.
Hayek, F. (1937). Economics and knowledge. *Economica, 4*, 33–54.
Hayes, M. (2010). *John Maynard Keynes*. Polity Press.
Hoover, K. (1997). Is there a place for rational expectations in Keynes's *General theory?* In G. C. Harcourt, & P. A. Riach (Eds.), *A "second edition" of the General theory*. Routledge.
Keynes, J. M. ([1921] 1973). *A treatise on probability. The collected writings of John Maynard Keynes* (Vol. VIII). Macmillan.
Keynes, J. M. ([1936] 1973). *The general theory of employment, interest and money. The collected writings of John Maynard Keynes* (Vol. VII). Macmillan.
Keynes, J. M. (1937). The general theory of employment. *Quarterly Journal of Economics, 51*, 209–223.
Keynes, J. M. ([1937–38] 1979). Letters to Hugh Townshend. In Id. *The general theory and after: A supplement. The collected writings of John Maynard Keynes* (Vol. XXIX). Macmillan.
Knight, F. H. ([1921] 1964). *Risk, uncertainty and profit*. Augustus M. Kelley.
Kyburg Jr., H. E., (2003). Are there degrees of belief? *Journal of Applied Logic, 1*, 139–149.
Langlois, R. (1994). Risk and uncertainty. In P. J. Boettke (Ed.), *The Elgar companion to Austrian economics*. Edward Elgar.
LeRoy, S. F., & Singell Jr., L. D. (1987). Knight on risk and uncertainty. *Journal of Political Economy, 95*, 394–406.
Luce, D., & Raiffa, H. (1957). *Games and decisions*. Wiley.
McCann Jr., C. R. (1994). *Probability foundations of economic theory*. Routledge.
O'Donnell, R. M. (1989). *Keynes: Philosophy, economics and politics*. Macmillan.
Raffaelli, T. (2006). Keynes and philosophers. In R. Backhouse, & B. Bateman (Eds.), *Cambridge companion to Keynes*. Cambridge University Press.
Runde, J. (1994). Keynesian uncertainty and liquidity preference. *Cambridge Journal of Economics, 18*, 129–144.
Runde, J. (2001). Chances and choices: A note on probability and beliefs in economic theory. In U. Maki (Ed.), *The economic world view: Studies in the ontology of economics*. Cambridge University Press.
Runde, J., & Mizuhara, S. (Eds.). (2003). *The philosophy of Keynes's economics: Probability, uncertainty and convention*. Routledge.
Russell, B. (1922). Review of *A treatise on probability* by J. M. Keynes. *The Mathematical Gazette, 11*, 119–125.
Skidelsky, R. (1992). *John Maynard Keynes: The economist as saviour, 1920–1937*. Macmillan.
Venn, J. (1876). *The logic of chance* (2nd ed.). Macmillan.
Vercelli, A. (2018). Weight of argument and liquidity preference. In S. Dow, J. Jespersen, & G. Tily (Eds.), *The General Theory and Keynes for the 21st Century*. Edward Elgar.
Zappia, C. (2015). Keynes on probability and decision: Evidence from the correspondence with Hugh Townshend. *History of Economic Ideas, 23*, 145–164.
Zappia, C. (2016). Whither Keynesian probability? Impolite techniques for decision-making. *European Journal of the History of Economic Thought, 23*, 835–862.

Chapter 4
Von Neumann, Morgenstern and Decision-Making Under Risk

4.1 Introduction

According to Frank Knight and John Maynard Keynes, risk and uncertainty are inherently contradictory concepts. As discussed in the previous chapter, both scholars emphasized the incongruity of analyzing uncertain situations as merely risky. They argued that the application of probabilistic reasoning, which was the prevailing approach to dealing with risk, could not be taken for granted in the case of uncertainty, indeed it may have been impossible. However, during the period when Knight and Keynes were producing their works, the analysis of decisions under conditions of risk was not as well systematized as their objections to the employment of expected utility would suggest.

The solution proposed by Daniel Bernoulli in the eighteenth century, to evaluate a choice with uncertain outcomes in terms of expected utility—an average of the utilities of the possible outcomes weighted by probabilities—rather than in terms of expected payoffs, was a well-established reference among marginalist economists by the end of the nineteenth century. But Bernoulli's idea was accepted only to the extent that it was consistent with the application of the principle of diminishing marginal utility, not as a general criterion of choice. As discussed in Chap. 2, faced with the difficulty of defending Bernoulli's criterion on the basis of the actual behavior of economic agents—who participated in unfair lotteries and did not buy insurance even though the insurance premium offered to them was fair—economists such as William Stanley Jevons and Alfred Marshall preferred not to question the principle of diminishing marginal utility. Outside of the domain of choice under conditions of certainty, the criterion of utility maximization was not considered a general criterion.

With the gradual evolution from a 'cardinal' interpretation of utility—which presupposes a utility that is somehow precisely measurable—to an 'ordinal' interpretation—where utility is used simply to represent the order of preference among the objects of choice—the matter became even more complicated. Even assuming that the decision to participate or not to participate in a lottery could be evaluated

C. Zappia, *Uncertainty in Economics*, Springer Studies in the History of Economic Thought, https://doi.org/10.1007/978-3-031-91221-4_4

in relation to marginal utilities, these utilities still had to be measured, as Bernoulli had done by assuming a logarithmic utility function, exhibiting decreasing marginal utility. But if utility is only an expression of a preference order, and thus any monotonically increasing transformation of a given utility function represents the same preferences as the initial function, then the calculation of marginal utilities loses its meaning.

The reasons for the employment of the notion of ordinal utility, along the lines indicated by Vilfredo Pareto and then systematized by John Hicks (1939), are various. They pertain both to the internal evolution of economic theory and to the influence of new methodological approaches imposed on science in general. Although we cannot go into it here, it is important to recall that the typical approach of economic science until the end of the nineteenth century basically presupposed the ability of the theorist to understand the workings of the human mind through introspection. In this sense, it was thought possible to refer to utility as a one-dimensional measure capable of explaining economic behavior; this measure was taken to be known not so much on the basis of observation, but somehow a priori. Utility was regarded as an immanent notion of which the economist was as aware as any other individual, and thus also of the economic agent being analyzed. In some authors more than others, this was accompanied by the hope that psychologists might one day actually offer a yardstick for its measurement.[1]

There have always been some critics of this approach among economists, such as the American institutionalists with whom Knight was confronted. However, the lack of a precise empirical reference for the concepts used, in particular of utility as a number, became a real weakness of economics in the face of the developments of logical positivism, which became the approach *par excellence* in all sciences at the beginning of the twentieth century. In those years, there was a shift toward an economic science in which the empirical adequacy of a theory could only be justified on the basis of precisely identified logical relations between variables. This implied an increasing focus on the property of logical consistency of its statements, which implied a redefinition of the foundations of the theory, starting with the role of utility.

This redefinition was therefore accompanied by the idea that utility should be measured by observing actual choices and not, as had been the case, by inferring optimal choices from the employment of a notion such as marginal utility, which came to be considered empirically unsound. In this sense, the marginal rate of substitution, on which Hicks based his analysis, could much better represent how much an individual with a given initial endowment of goods would value an additional unit of a given good. The increase in utility resulting from an increase of one unit of good 1 in the consumption basket could be expressed in units of good 2, the units that the

[1] In particular, Francis Ysidro Edgeworth, who followed Jevons in conceiving of economic theory as a calculus of pleasure and pain, imagined the possibility of constructing a 'hedonimeter,' an instrument capable of measuring the pleasure of an individual in the same way that physical instruments measure physical magnitudes. Although mostly mentioned for the sake of argument, the hedonimeter would provide a direct measure of utility just as other sensations are measured, that is, by using the "just perceivable increment" of the sensation of pleasure as the unit of measurement (Moscati, 2018, pp. 53–54).

individual agent was (at most) willing to give in exchange, thus holding total utility constant. The change could then be evaluated in terms of a number with a precise meaning and, for marginal increases in good 1, could be represented as the slope of an indifference curve passing through the initial endowment.

But the decisive step toward a complete abandonment of the traditional use of the concept of utility was the theory of revealed preferences proposed by Paul Samuelson, a Harvard economist who would become a dominant figure of post-World War II economics after joining the economics department of the Massachusetts Institute of Technology. Samuelson (1938) argued that the only real foundational element of choice theory had to be an assessment of the consistency, or otherwise, of the observed choices. A traditional utility function could still be used, but only on the basis of the preliminary step of verifying the consistency of the individual's demand for goods at given prices. That is, verifying the fact that if an individual chooses basket *A* when basket *B* can be chosen, at given prices, then the same individual can choose *B* at different prices only when *A* cannot be purchased at those prices—according to the so-called (weak) axiom of revealed preferences.

It was within this broader context that even in economics formal rigor began to supplant the simple mathematical representation of concepts. The seminal volume *Theory of Games and Economic Behavior* by John von Neumann and Oskar Morgenstern, published in 1944, is unanimously acknowledged as the main example of this new method of analysis. It is to the *Theory of Games* that we owe, as a joint product of the project to analyze strategic situations in economics in an axiomatic context, a redefinition of the concept of measurable utility and the consequent definition of the expected utility function for decisions under risk, which is still commonly used today.

4.2 Von Neumann and Morgenstern's *Theory of Games*

The Hungarian mathematician John von Neumann and the Austrian economist Oskar Morgenstern (whom we will abbreviate as vNM when quoted together) combined their efforts in the production of the *Theory of Games* after they had already emigrated to North-American soil, both having fled continental Europe in the years of the spread of Nazism and anti-Semitism, and both coincidentally affiliated with Princeton University.

Von Neumann (1903–1957), at birth Janos but later called 'Johnny' by his American colleagues, is unanimously recognized as one of the greatest mathematicians of the twentieth century, a major participant in the axiomatic revolution that spread from geometry to all the exact sciences. In the axiomatic approach—inspired by the work on the foundations of geometry by David Hilbert, with whom von Neumann collaborated in the early 1920s—the principle holds that only on the basis of precise assumptions expressed in mathematical language (the axioms) can cognitively useful conclusions be consistently derived. Within the logical positivist program, the axiomatization of a theory is a crucial step because it enables the identification of the theoretical terms for a correct understanding of the observational consequences of a theory, concepts and explanations that organize the empirical knowledge of the world (Boumans & Davis, 2010). However, empirical verification of the theory was only the second step in theoretical reasoning, although it was on the basis of the 'verifiability' of the results through comparison with real data that truly scientific theories could be distinguished from metaphysical speculations.[2]

The emblem of this 'formalist' revolution in economics is considered to be von Neumann's essay entitled "A model of general economic equilibrium." The essay was initially published in German in the proceedings of the Vienna Colloquium, a periodic seminar run by the mathematician Karl Menger where mathematicians, physicists, philosophers and economists discussed the new formal tools suggested by Hilbert's approach. Von Neumann's essay is characterized by the use of some demonstrative techniques that would form the basis of the most formalized areas of economics in the years to come, both the theory of general economic equilibrium and the theory of growth. The approach taken by von Neumann in this essay would become commonplace among mathematical economists in the years to follow, with the subject matter briefly identified in the introductory remarks, a set of assumptions clearly put in evidence as mathematical relations among precisely defined variables, and a result identified through a theorem, that is, a mathematical proof. Von Neumann argued that, when analyzing complex economic systems, results such as that of the existence of a price vector that solves a general competitive equilibrium of goods markets "*cannot be proved by a qualitative argument*," but require a proof that can only be obtained by using mathematical tools that economists were unfamiliar with (von Neumann, [1937] 1945, p. 1, original italics).

Among the many interests of von Neumann—who would later be involved in the development of the atomic bomb known as the Manhattan Project and in foundational research on digital computing—was the study of strategic interactions between individuals. Von Neumann had already approached the analysis of strategic behavior by proposing a purely mathematical representation of games such as chess in a 1928 article in German entitled "On the theory of games of strategy." The article is considered the first accomplished exposition of what would later be called non-cooperative zero-sum games, as well as the crucial concept for their solution, known as 'max-min' equilibrium. Von Neumann ([1928] 1959) believed that interactions between individuals and between groups, both competitive and coalition-generating (later analyzed under the heading of cooperative games), were central themes for any meaningful

[2] Following Karl Popper's analysis, the criterion of the 'verifiability' of a theory was soon replaced by that of 'falsifiability' (Popper, [1934] 1959). Obviously, the results to which a theory leads must be confronted with real data, but with Popper the illusory ambition to verify a general proposition in all cases to which it may apply was abandoned. Rather, the focus became the possibility that verification can be carried out in general, that is, that the theory has a logical structure such that it can actually be verified. And if it can be proven in even one relevant case that it is not, then it is deemed falsified, and must either be amended or abandoned. According to Popper, if this process of falsification cannot be undertaken, one is not dealing with theories, but only with metaphysical propositions (Boumans & Davis, 2010, pp. 74–76).

analysis of social systems. But he was also convinced that economic theory did not adequately address them.

Oskar Morgenstern (1902–1977) was an economist with a keen interest in forecasting, a subject to which he devoted himself from an applied point of view as director of the Austrian Institute for Business Cycle Research in Vienna in the early 1930s. In his more speculative analysis of the subject, he dealt with the relationship of time and foresight to general equilibrium theory. He was a critic of the assumption of 'perfect foresight,' that he believed was common in the economic theory at the time. Morgenstern argued that in dynamic contexts, that is, when trying to represent the effect of the passage of time on economic variables, the analysis must necessarily focus on the expectations of agents. But to assume that correct expectations can be a precondition for the realization of equilibrium, as was typical of the equilibrium analyses of his contemporaries, he deemed inappropriate.

In a famous example, later re-examined in the *Theory of Games*, he analyzed the expectations that guide the behavior of Sherlock Holmes and his adversary Moriarty—when one has to make a decision without knowing what the other's choice will be. Morgenstern ([1935] 1976, pp. 173–174) used the example with the aim to show that an expectation about the behavior of another individual can generate an unbroken chain of reciprocal assumptions about the behavior of others, to the point where it becomes impossible to realize an equilibrium of choices.[3] An assumption such as that of perfect competition, in which individual agents are so 'small' relative to the market that their choices are irrelevant to the determination of prices, which they can then take as given, is adopted by economists precisely in order to avoid dealing with the kinds of problems that characterize strategic interaction. And even when the strategic environment of a market was analyzed, such as in oligopoly theory, Morgenstern concluded, it was done with inadequate mathematical tools.

Morgenstern therefore criticized economic theory for having a biased view of the problems related to the actual knowledge that individuals have of the problems they face. It should be noted that this criticism is reminiscent of both Knight's criticism of perfect competition as a static theory and Keynes's criticism of 'classical' economists for their treating of the future as if it were known. But Morgenstern adopted a different viewpoint from those of Knight and Keynes. When still in Vienna, Morgenstern actively participated in Menger's Mathematical Colloquium, and derived the conviction that the inability of economic theory to achieve agreed and empirically relevant results was primarily due to the inaccuracy of the type of

---

[3] The situation imagined by Morgenstern, taken from a Conan Doyle's novel, was that of Sherlock Holmes wishing to reach Dover in order to embark for the continent and escape Professor Moriarty who was chasing him. Upon boarding the train in London, Holmes sees Moriarty appear on the tracks and imagines that he is taking a fast train to get to Dover first. The choice facing Holmes is whether to go to Dover or, rather, to get off at the only intermediate stop, Canterbury. His opponent, who is at least as clever, must make a similar choice, and both must choose what to do without knowing what the other will do. If the result of the choices is that they arrive at the same stop, Holmes is killed, otherwise he is saved. The structure is that of a game that, in the *Theory of Games*, would be referred to as a zero-sum game, because the gain (or loss) of one player is matched by an equal loss (or gain) of the other player.

deductive reasoning adopted. Under Menger's influence, at the Colloquium it was agreed that the logical-deductive approach of economists should follow the approach proposed for the exact sciences, acquiring the formal precision guaranteed by axiomatization. This approach Morgenstern succeeded in developing after he emigrated to the United States and met von Neumann.[4]

It was largely at Morgenstern's urging that von Neumann, after their meeting at Princeton University, returned to the themes of the 1928 article on games and began to work on their possible generalization to social contexts.[5] The meeting between the two scholars, who were both interested in the subject of the interaction between individuals, resulted in the contribution that gave rise to a mathematical theory of strategic situations that was intended to be applied to economics, but which would prove influential in many other fields. The importance of the *Theory of Games* lies both in its method and in its actual content. As regards the method, it is the axiomatic approach, used for the first time in an accomplished manner in economics, according to which the relevant results of a theory are only those that can be deduced through theorems on the basis of initial assumptions, which must be formally stated propositions. As regards the content, it is the extension of economic theory beyond the simplifications allowed by the assumption of perfect competition to the analysis of actual interactions between individuals.

However, the aspect that interests us here is that the *Theory of Games* also contains a specific representation of the problem of choice under conditions of risk, a representation that would become the prevailing theory. This is a version of expected utility theory that enabled the overcoming of Marshall's criticism of the non-axiomatized version provided by Bernoulli.

4.3 Expected Utility and Decisions Under Risk

In the introductory chapter of their *Theory of Games*, vNM argued that an analysis of the interactions in economic systems cannot be based on the choices of an isolated individual agent *à la* Robinson Crusoe, who solves a maximum problem given external constraints on his actions. In the interaction between several individuals, which characterizes a "social exchange economy," each individual must take

[4] Interestingly, it was precisely because of this methodological requirement that Morgenstern was highly critical of Keynes's *General Theory*. He approved of Keynes's aim to put the issue of expectations at the center of his analysis, but found it "so vague that I think we shall have to wait for further elucidation from his pen before making a final judgement." Indeed, Morgenstern (1937, pp. 158–159) continued, "it is not enough simply to refer to expectations … we need to know what they are determined by, what factors they depend on, and the ways in which they are mutually interdependent. Keynes offers no real analysis of these points."

[5] On the genesis of the *Theory of Games*, see Leonard (1995). Mirowski (2002) and Giocoli (2003) provide a particularly detailed, though not necessarily concordant, analysis of what kind of game theory von Neumann actually had in mind, placing emphasis on the differences with the widely accepted version attributed to John Nash, as illustrated for instance in Myerson (1999).

into account the behavior of other individuals. Thus, among the variables that characterize the external constraints, there are at least some that cannot be taken into account as known at the time of the decision, and these are, precisely, the choices of other individuals. The consequence is that the individual maximum problem is no longer properly defined, but it is a matter of solving a "peculiar and disconcerting mixture of several conflicting maximum problems" (von Neumann & Morgenstern, [1947] 2004, p. 11).

However, evaluating the outcome of an individual's own actions when they are influenced by the individual's expectations of the actions of other individuals, which in turn reflect the others' expectations on the individual's actions, is such a complicated problem that it requires simplifications. Traditional economic theory should be criticized because, as Morgenstern had already pointed out in his early work, the simplification it adopts—that of assuming that the usual form of the market is perfect competition—removes rather than solves the problem. Instead, vNM proposed to start with simple interaction problems, those involving only two individuals, and then to generalize the analysis to more and more individuals, up to n individuals. All problems that are tractable within the same general scheme of reasoning would then be considered exemplifications of that scheme, but the scheme must be an axiomatized formal structure representing "the mathematically complete principles which define 'rational behavior' for the participants in a social economy" (von Neumann & Morgenstern, [1947] 2004, p. 31). Hence the analysis of vNM's games that later became conventional in economic theory, where the description of the strategic situation is conducted in terms of players, moves (in sequential games) and strategies (in normal-form games), and payoffs for each player. Having represented the strategic situation in this way, the next step was to identify which of the outcomes could be expected to be the outcome of the strategic interaction, that is, the 'solution' of the game.

Still, the characterization of the individual's behavior required a preliminary step. The results of any economic activity, and so of the interaction that the theoretician wishes to study by means of a game, must be evaluated in relation to the utility that the individuals derive from it. And this must be done in a context in which—as economists almost unanimously agreed after Pareto—utility is ordinal, and therefore must be non-numerical in nature, in order to be plausible. In a retrospective article on the collaboration with von Neumann, Morgenstern (1976) reported that in the first meetings with von Neumann the problem of the unit of measurement of the payoffs obtained by the players as a result of their choices immediately arose. These outcomes appear, for example, as the numbers in the matrices with which they intended to represent the games in normal form, those made classic by the so-called prisoner's dilemma (Luce & Raiffa, 1957, pp. 94–97).

The economist Morgenstern recalled insisting with the mathematician von Neumann that these numbers could not represent money, but had to represent utilities; and that if utilities were to be numbers, this had to be justified in some way, otherwise economists would not accept the representation. Morgenstern recalled that both he and von Neumann were well aware of Karl Menger's article on the St. Petersburg Paradox, in which Menger ([1934] 1979) had presented in detail the employment of the expected utility function proposed by Bernoulli for risk. But even Menger, a prominent mathematician, had not considered the axiomatic foundations necessary for the employment of a numerical utility. When confronted with this problem, von Neumann solved it in a way that he considered so simple that he was surprised that no one had thought of it before him. It can be shown, vNM wrote in the *Theory of Games*, that "under the conditions on which the indifference curves analysis is based, very little extra effort is needed to reach a numerical utility" (von Neumann & Morgenstern, [1947] 2004], p. 17).[6]

The theory of choice under conditions of certainty underlying the representation of preferences by indifference curves assumes that individuals have complete and transitive preferences, that they are always able to rank two different objects—which in the theory of consumption are baskets of goods—without falling into contradiction. This must happen, von Neumann argued, even when the objects of choice include those typical of uncertain situations, that is, when the choice concerns two possible future outcomes that are alternative to each other. The further assumption that individuals know how to rank combinations of these objects with the probabilities with which they can occur, seemed to him a "natural extension" of choice theory. In fact, this is implicitly the case for all those economic activities whose outcome depends on the probabilities of events over which individuals have no control, such as the weather in agriculture. For the sake of simplicity, however, these objects were thought of as lotteries. The solution proposed by von Neumann was to include lotteries among the objects of choice—objects that are combinations of other objects with specified probabilities, also called prospects—and then to deduce utilities from the choices on these mixed objects, as numbers representing the preferences expressed in the choice.

To illustrate, von Neumann and Morgenstern ([1947] 2004, p. 18) used as reference three alternatives, the objects of choice A, B and C, and a preference structure possessed by an individual such that, for instance, $A \succ B \succ$ C. An individual is assumed to possess a "clear intuition" also when examining combinations of these alternatives, which are therefore assumed to be orderable as well. Just as it is usually assumed that a person knows how to order B with respect to A and C, it is also assumed that she is able to order B with respect to a combination of A and C. For instance, she is assumed to be able to say whether B is preferred, indifferent or not preferred with respect to object A with probability ½ and C with probability ½. A and C are the usual objects of choice under conditions of certainty, but all their probability mixtures now become objects of choice, and can be compared with B. The combination that with given probabilities, α and $1 - \alpha$, ensures the outcome A or C can be thought of as a lottery, L. As in any lottery, the outcomes A and C are alternatives, in the sense that if one is realized, the other cannot be realized.

[6] Actually, as we shall see in the next chapter, von Neumann's solution had already been formulated by Frank Ramsey when he dealt with the notion of subjective probability. Ramsey's 1926 essay on probability became well known in the 1950s, after Leonard Savage's axiomatization of subjective expected utilities. But it had remained little known outside Cambridge for a long time.

VNM pointed out that, by reasoning in this way, important new information about the person's preferences can be obtained. Indeed, at this point, it makes perfect sense to deduce from the person's choices what are the probabilities that make *B* indifferent to *L* as a (probability) mixture of *A* and *C*. For example, if it is true that *B* is considered to be indifferent to the lottery that gives *A* with probability 2/3 and *C* with probability 1/3, then for the person must be $u(B) = (2/3)u(A) + (1/3)u(C)$.[7] Without loss of generality, it can be assumed that in an ordered set the alternative to which all others are preferred has utility 0, and the alternative that has no preferred alternatives has utility 1: in the three-alternative example we used, it can be assumed that $u(A) = 1$ and $u(C) = 0$. But then $u(B) = 2/3$.

Assuming that the choices involve not only certain alternatives but also risky alternatives—which we have called lotteries—it is possible to repeat the reasoning just carried out with combinations of other objects, and thus to derive from the stated choices a utility function over the entire domain of choice. Moreover, this function is numerical, in the sense that utility is actually measured by means of probabilities; specifically, the derived utility function is a probability-weighted average of utilities. In general, if a lottery is represented by a set of possible outcomes, *x*, and associated known probabilities, *p*, then the lottery is valued according to $\Sigma_i\ p_i\ u(x_i)$. The name expected utility derives from this characteristic, and the specific procedure used to derive it makes it a 'vNM utility function,' as it is still called.[8]

VNM emphasized that they did not postulate the existence of a traditional numerical, and thus cardinal, utility; rather, they postulated that it was meaningful to combine outcomes with probabilities. However, given that a utility function can be derived in this way, there is no reason to assume that individuals would not maximize this function when faced with decisions about such mixed objects, that is, lotteries. Utilities are derived from choices, which are in principle observable, and can then be used to apply a decision criterion. Rational choice under risk can thus be represented as economists have always done for choice under certainty, as the result of a maximization, in this case of an expected utility function.

To summarize, VNM's way of arguing was typical of what later came to be called a 'representation theorem:' if a person is able to express preferences between each

---

[7] Actually, the assumption that the individual always has an indifferent combination of *A* and *C* for each situation *B* is an axiom of the theory, called the 'continuity axiom.' In addition to the other usual assumptions under conditions of certainty, such as completeness and transitivity of preferences, it is assumed that for each *A* and *C* there always exists a lottery *L*, made up of a mixture of *A* and *C* with probabilities α and $1 - \alpha$, that is indifferent to any *B*. This 'technical' assumption is necessary to derive the numerical function representing the ordering between the lotteries (Moscati, 2023, pp. 27–28).

[8] It should also be noted that since the values of the extreme alternatives can be set arbitrarily, any other value of $u(A)$ and $u(C)$ that delimits the range of utilities is permissible, e.g., $u(A) = 100$ and $u(C) = 0$. But this simply means that the utility function derived from the example is not unique, or rather, in mathematical language, that it is unique up to an increasing linear transformation. In fact, proceeding as in the text, it follows from these new values that $u(B) = (2/3)100$. But the differences between the utilities have the same relative relationship as in the previous case, because the utility function is cardinal. This is what makes a linear transformation different from a monotonic transformation.

possible pair of lotteries, where the lotteries are derived from a given set of basic alternatives, then it is possible to introduce utility values associated with the basic alternatives in such a way that, if the person is guided solely by the search for the maximum expected utility value, she will choose in accordance with her preferences. Put into words, this is the fundamental theorem of all modern decision theory.[9]

We have already mentioned that Bernoulli had introduced an expected utility procedure, and that Marshall had restricted its use. But the starting assumption of Bernoulli's and Marshall's reasoning was utility, not choices: given a specific utility function, choices were determined as the outcome of a maximization problem. And since marginal utility could only be decreasing, these choices had to be evaluated on the basis of decreasing marginal utility. Hence Marshall's conclusion that the observation of many choices under risky conditions—such as buying a lottery at its expected value—clearly disproved the possibility of representing individuals as if they were using the criterion of maximizing expected utility. In vNM, on the contrary, it is utility that is derived from choices. Therefore, the criterion of choosing by maximizing the expected utility is valid in general, regardless of the kind of utility function that is revealed by choices. As pointed out first in Friedman and Savage (1952) and, even more clearly, in Ellsberg (1954), it is as if we are talking about two different concepts, since the expected utility of vNM, although cardinal, is still only an index for making comparisons, as is the ordinal utility of Pareto.[10]

Precisely because it was directly related to a traditional topic of discussion among economists, the meaning of the newly introduced expected utility function initially remained the most discussed of the various contributions of the *Theory of Games*. However, before addressing some crucial aspects of this discussion, it is necessary to emphasize the essentially instrumental use that vNM made of the utility function, since their aim was to proceed in the analysis of games with outcomes that could be represented numerically. The axiomatic derivation of expected utility was obviously necessary, given the general framework of their work. But it was only announced in the first edition of the *Theory of Games*, and when it was actually elaborated, in an appendix added to the 1947 edition, it was extremely complex. The appendix demonstrated the fundamental theorem according to which compliance with a set of axioms was shown to be equivalent to the maximization of an expected utility function. But the proof was a rather convoluted mathematical demonstration in which the crucial axiom for the demonstration was formulated in an obscure way, if not to mathematicians surely to contemporary economists. It was the scholars of decision theory who would clarify that, in addition to the axioms of completeness and transitivity underlying the standard construction of indifference curves for choice under

---

[9] In decision theory, a 'representation theorem' proves the existence of an isomorphism between a class of preferences and a class of numerical functions. VNM proved that if the decision-maker's preferences satisfy a certain set of axioms, then these preferences can be represented by a numerical (expected) utility function. That is, if $L_1 \succ L_2$ then $\mathrm{E}u(L_1) > \mathrm{E}u(L_2)$, and vice versa.

[10] Indeed, a correct notation should distinguish the utility of the objects of choice from the expected utility of lotteries, using two different symbols, one for Bernoulli's utility function and the other for vNM's (expected) utility function (see Moscati, 2023, pp. 31–32).

conditions of certainty, vNM were substantially using a crucial new axiom, which came to be called the 'independence axiom.'

The independence axiom can be formulated as follows. If there are three lotteries among the objects of choice, L_1, L_2, L_3, and if it is true that an individual's preferences are such that $L_1 \succ L_2$, then the combination of L_1 with L_3, with probability α and $1 - \alpha$, must be preferred to the combination of L_2 with L_3, with probability α and $1 - \alpha$. This formulation is due to Jacob Marschak, a mathematical economist working at the Cowles Commission for Research in Economics—a research center affiliated with the University of Chicago that was to play a key role in those years in the development of a number of formal techniques that would later form the basis of, among others, activity analysis and econometrics (Dimand, 2020). With the reinterpretation of the axioms, and in particular with the identification of the axiom of independence, Marschak (1950) endeavored to convince economists to accept the new approach: the axioms of vNM were plausible as an ideal model of human preferences, and thus applicable outside the context of game theory.[11]

The independence axiom rationalizes a consideration that vNM had made in their introductory chapter, but then had not made explicit in the formal proof of the theorem, which was set up in terms of a methodology—indifference classes between lotteries—that was less immediate. The idea was that adding an element that might occur with a given probability to lotteries between which an order of preference has already been established, cannot change the original preference formulated when that element had not been considered. According to von Neumann, as we have already noted in quoting Morgenstern's account, this was evident because in lotteries either one outcome occurs or another outcome occurs, and the outcomes are not alternatives in the traditional sense of choice under conditions of certainty, but are mutually exclusive. Under certainty, on the contrary, if a person chooses in a given way, for example, between the goods 'water' and 'wine' when she does not have the possibility of consuming the good 'food,' it cannot be assumed that the stated preference would remain unchanged if she also has the possibility of consuming 'food.' The axiom of independence thus implies that there is neither complementarity nor substitutability in preferences, but this assumption was considered inapplicable by the majority of economists at the time. While von Neumann found it plausible for choices in a probabilistic context, such as when dealing with lotteries, its use was initially opposed by many, including Samuelson himself, who was known precisely for his studies in the field of choice theory.[12]

---

[11] Marschak (1946, p. 114) was one of the earliest and most enthusiastic reviewers of the *Theory of Games*, but he argued that the book's greatest contribution laid more in the "tools of modern logic" that it suggested could be applied to economics, than in its concrete results. VNM's axiomatic method constituted an antidote to "any subconscious smuggling-in of undefined terms and operations or of assertions that have not been proved."

[12] The debate over the independence axiom will play a crucial role in the years to follow, both for the acceptance of the method of analysis and for its possible application beyond the context of risk with objectively given probabilities (Fishburn & Wakker, 1995). Moscati (2016) provides an assessment of Samuelson's changing attitude toward the axiom.

More generally, it will be the task of economists interested in decision theory to clarify the boundaries of expected utility theory and indicate its possible applications. Milton Friedman and Leonard Savage, in particular, explained how the traditional concerns about the employment of utility maximization outside of choice under certainty could be overcome. It may happen that individuals participate in unfair lotteries, or do not buy insurance when offered at a fair insurance premium, because they have different attitudes toward risk. But these different attitudes are all compatible with the maximization of a utility function, if it is the expected utility function defined by vNM. Friedman and Savage (1948, pp. 288–291) introduced the simplifying methodology of considering the objects of choice as monetary incomes, and then showed how it is possible to study an income utility function justifying both a risk-averse and a risk-seeking attitude. If the combination of A and C with probability α and $1 - \alpha$ is the lottery L with expected utility $u(L) = \alpha u(A) + (1 - \alpha)u(C)$, it is possible to compare this utility with the utility of the expected value of the lottery, $u[\alpha A + (1 - \alpha)C]$. A value of $u[\alpha A + (1 - \alpha)C]$ greater than $u(L)$ signals 'risk aversion,' which can also be measured in money if A and C are monetary incomes.

From this, Friedman and Savage derived the definition of a 'risk premium' that is still in use today. Since the expected monetary value of the lottery is its fair value, if $u(L) < u[\alpha A + (1 - \alpha)C]$, the difference between the two utilities—which is now a number—can be seen as a measure of how risky the subject judges the lottery to be. Furthermore, if this difference can be measured on the axis of utility, the same can be done on the axis of income, so that the risk premium can be measured in money, which can also be seen as indicating the maximum amount a person is willing to pay to avoid the risk of participating in the lottery (see Fig. 4.1, which is reproduced from the original paper, with the variable I indicating money income). Conversely, a person who agrees to participate in the lottery can be regarded as a 'risk-loving' person, since she values $u(L) > u[\alpha A + (1 - \alpha)C]$.[13]

The use of lotteries as an example for decision-making under risk would later become commonplace. Friedman and Savage (1948, p. 287) noted that lotteries "seem to be an extremely fruitful, and much neglected, source of information about reactions of individuals to risk," a source presenting risk in a "relatively pure form, with little admixture of other factors." They pointed out that studies about lotteries had always been conducted throughout history, providing very extensive evidence of the phenomenon. But they also noted that lotteries lend themselves to being used as objects of "experimentation:" it is easy to study the choices of individuals who are asked to make decisions in controlled contexts, and to derive regularities of behavior

[13] Let us return to Bernoulli's case from Chap. 2, the case of the poor individual who finds a lottery ticket on the street and has to choose whether to sell it for 10,000 ducats, or to keep it and participate in the lottery, which can yield 0 or 20,000 with probabilities ½ and ½. By following vNM's procedure, it is now possible to reconstruct from the poor's choices, first, what minimum amount of money he will accept in return from a rich person who wants to buy his ticket and, second, his entire expected utility function. Friedman and Savage designated "equivalent income" as the sure amount I such that $u(I) = u(L)$, where in this instance $u(L) = (1/2)u(20{,}000) + (1/2)u(0)$. This is evidently what modern microeconomics textbooks refer to as the certainty equivalent of the lottery, a measure that economists had often referred to prior to vNM, albeit only intuitively.

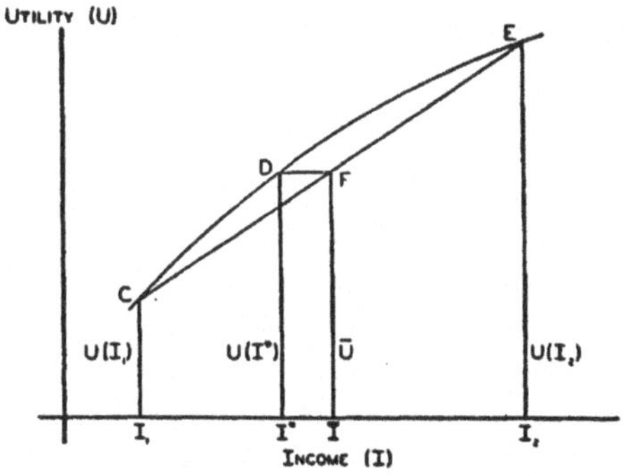

Fig. 4.1 Equivalent income and risk premium (reproduced from Friedman & Savage, 1948, p. 290)

from the observed choices. That of vNM is thus a "hypothesis [that] has implications for behavior … that are capable to be contradicted by observable data" (Friedman & Savage, 1948, p. 304) in deference to the scientific method of the exact sciences.

On the basis of the empirical work constructed in this way by Mosteller and Nogee (1951), in which the theory of expected utility was not falsified, Friedman and Savage (1952, p. 466) added that it was already possible to confirm that the theory "must be regarded as a promising conjecture."[14] However, they argued that the main appeal of vNM's hypothesis derived from "indirect evidence rendering it plausible," that is, evidence provided "by the coherence of the hypothesis with the rest of economic theory." Friedman and Savage were reasoning on the question of whether the theory of expected utility should be interpreted as a true description of the behavior of individuals, or whether it was not instead only a theoretical contribution of a normative nature. Mostly in favor of the second interpretation, they suggested that "further work on the postulational base of the hypothesis" was under way, and presented a first version of Savage's own axiomatization, aimed at including into expected utility theory gambles on any kind of objects, not only lotteries. The debate on this question—whether decision theory has a descriptive or a normative character—opened in those very years, but continues to this day, as we shall see in the following chapters.

[14] Mosteller and Nogee (1951) implemented the strategy suggested by Friedman and Savage in a seminal controlled laboratory experiment with Ph.D. students. They inferred utility measures from the students' choices regarding a first set of lotteries, and then used these utilities to predict the students' choices regarding a second set of lotteries. Their conclusion was that it was not unreasonable to assume that decision-makers behave in such a way to maximize their expected utility. The story of the experimental measurement of utility is recounted in Moscati (2018).

4.4 Probability and Uncertainty in Von Neumann and Morgenstern

As mentioned above, expected utility theory was discussed extensively in those years, but not only for the formal technical aspects highlighted by Friedman and Savage. There are at least two other related issues of interest for a history of uncertainty in economics. A first issue is of course that of the notion of probability used in the *Theory of Games*. In the aforementioned article on the collaboration with von Neumann, Morgenstern (1976, p. 809) recalled that they were of course aware that introducing probability in decision-making might have meant dealing with a slippery subject such as the logical foundations of the concept. This was the reason why they decided to base their argument on a frequentist view of probability, assuming a notion of probability that was 'objective' knowledge on the part of individuals.

In the introductory chapter of their volume, vNM admitted that "probability has often been visualized as a subjective concept more or less in the nature of an estimate," evidently referring back to Knight, with whom Morgenstern had been in contact for the English translation of his 1935 article criticizing perfect foresight. Yet, for their goal of constructing an individual measure of utility, the simplest procedure was to separate the question of subjective utility from that of probability. Had they had to consider also probability as subjective, they would have had to give it an axiomatic foundation. Thus, they took as their reference "the alternative and perfectly well-founded interpretation of probability as frequency in long runs," that is, an objective notion of probability (von Neumann & Morgenstern, 1947, p. 19). Of course, the question of the extension of the analysis to a subjective probabilistic context could not be abandoned by decision theorists, and indeed it constitutes the main topic of discussion in the following years. But it is on the basis of this view—of an uncertainty limited in effect to the risk faced in lottery-like situations—that combinations of objects of choice were considered "natural" by vNM.[15]

The second question is linked to the previous one, because it still concerns probability, but it relates to the sense in which probability is used within game theory. As we have seen, a game is a situation in which the payoffs of a player are uncertain due to the presence of the other players, and to the fact that the choice of the others is unknown when the individual player has to make his decision. However, the representation of the game in terms of players, strategies, payoffs and utilities associated with payoffs, is supposed to be known to all players, so that each player knows what the other players' possible payoffs are. Furthermore, each player assumes that the other players choose according to a criterion that coincides with their own, that is, to obtain the maximum possible outcome for themselves.

[15] VNM mentioned in a footnote that they had considered whether to axiomatize utility and probability at the same time, in a purely subjective setting, but this would only have complicated the matter. Evidently, this choice limited the field of investigation to situations that, after Knight, were considered to be of risk. But given the instrumental character of utility theory in relation to the main objective of their work—a general theory of strategic behavior—they decided to proceed otherwise.

Von Neumann's idea for solving a game, the one already presented in the 1928 article, was to neglect the cognitive aspects of the strategic situation and to imagine that the player chooses in such a way as to limit as much as possible the uncertainty associated with the possible choices of the other players. Faced with a situation of complex interaction, von Neumann simplified it by hypothesizing as a criterion of choice that of choosing the strategy that maximizes the minimum payoff associated with each of one's own possible strategies, depending on the choices of the other. In a two-person game, an individual knows that the outcome he will obtain from each strategy he chooses is determined by the choice the other is making at the same time. Therefore, von Neumann imagined that the subject reacts to the situation of uncertainty by focusing only on the worst outcome associated with each of his own possible strategies, given the other's choice. In other words, after identifying the worst possible payoff for each of his own strategies, the chosen strategy is supposed to be the one that yields the highest of the lowest possible payoffs, a strategy therefore called the max-min strategy. In principle, this reasoning is attributed to all players, and all players assume that the others are doing the same.

In the case of the simplest non-cooperative game—the zero-sum game between two players, that is, where a positive payoff of one player corresponds to a negative of the same magnitude for the other player, and vice versa—the solution that von Neumann identified was that in which while one subject chooses to maximize the minimum possible payoff (gain), the other subject, using the same reasoning, can be considered as choosing the strategy that minimizes his maximum possible payoff (loss). In fact, since the first player's winnings are the second player's losses, the strategy chosen by the second player will be the one that yields the smallest possible outcome, thus called the min-max strategy. By choosing this strategy, each player eliminates the uncertainty inherent in the opponent's behavior, in the sense that a minimum level of utility is ensured, either because it maximizes the min or because it minimizes the max.

Contrary to what one might think of a typical strategic situation, in this way there is no need to reason about the probability with which the other player will make one choice rather than another. In fact, even in the case that had interested Morgenstern before meeting von Neumann, that of Holmes and Moriarty illustrated earlier, there is a way of solving the question of the chain of reasoning that initially seemed unsolvable to Morgenstern because of the infinite regress—I know that he knows that I know …, etc. By disregarding the question of who of the two players is the better 'psychologist' and concentrating instead on what, according to von Neumann, is strategically relevant from an objective point of view, one obtains a solution to the game.[16] In other words, those used by the statistician Abraham Wald (1945), one

[16] The importance of von Neumann's reasoning is, of course, also linked to a mathematical aspect: von Neumann proved a theorem that defines the conditions ensuring that, while one player chooses the max-min strategy and the other a min-max strategy, an equilibrium of choices emerges as the solution to the game. That is, the max-min/min-max behavior warrants the existence of an interpersonal kind of consistency: if both players follow this rule, their strategies are mutually consistent and neither player has any reason to change his behavior, so the game is indeed solved. But this interpretation of the game would be contradicted by the developments related to John

can think of the situation as if each individual believes that he is playing against a malevolent 'Mother Nature' that always chooses with the aim of harming him, so that the optimal choice can only be to focus on the maximum loss that Nature can inflict on him for each of his strategies, and then to choose the strategy that minimizes the maximum possible loss.[17]

The max-min strategy (or min-max, depending on the player taken as reference) expresses an extremely precautionary approach, basically a pessimistic view of the possible outcomes of the strategic situation. Moreover, the solution proposed by vNM was not the result of a real analysis of why it is rational to be pessimistic. It is suggested only because it provides a way to limit strategic uncertainty. Why one should accept this solution, and what would be the extreme consequences of this way of reasoning, became a question soon debated by decision theorists. Daniel Ellsberg, a young Harvard graduate with a Master's thesis on game theory, objected that it seemed unjustified to him not to reason problematically about the behavior attributed to others in a context in which the decisions of others are crucial to the outcome of the game.

Ellsberg (1956, p. 914) argued that the strategy of maximizing the first player's minimum possible payoff makes sense to the player because it is based on the conjecture that the second player is guided by the same principle of behavior thought of by von Neumann, and that he knows that the other is making the same conjecture. But in a strategic context it is not evident that assuming that the other follows a certain kind of behavior is the best way to act, if only because the assumption about the other's behavior cannot be verified until the end of the game. Ellsberg observed—as he would later do in his famous work on uncertainty, which we will discuss in Chap. 6—that it is one thing to identify a reasonable way of making decisions, and quite another to assume that this is the *only* reasonable—therefore rational—way of doing so. We will return to this argument of Ellsberg's later, but it is interesting to note that he believed that vNM did not really want to deal with the uncertainty inherent in the strategic situation: "Uncertainty is a state of mind, a property of belief or expectation; if it is present it cannot simply be 'assumed away'" (Ellsberg, 1956, p. 914). This contention gave rise to a way of looking at strategic situations that, while starting from vNM, would later lead to a method of analysis quite different from theirs, according to what, for reasons that we will examine in the next chapters, is considered a 'Bayesian' approach.[18]

Nash and his idea of the solution as an equilibrium of expectations (Giocoli, 2003). For a simple exposition of the steps to identify the solution of a zero-sum game according to vNM—the so-called minimax theorem—see Luce and Raiffa (1957, Ch. 4). Luce and Raiffa also illustrates the sense in which Nash's solution includes the minimax as a special case.

[17] We cannot elaborate on Wald's role here, but it is nonetheless important to note that Wald's idea that the statistician attempting to derive the value of a stochastic variable from a set of data must imagine that he is solving a decision problem under conditions of probabilistic risk, and thus must minimize the risk of error in estimating the variable, underlies the developments in decision theory in those years (Giocoli, 2013).

[18] Also, the idea of a mixed strategy, which vNM used to solve the problems that arise when a zero-sum game has no solution, does not use probability in the sense we are wondering about here,

4.5 Expected Utility and Portfolio Choices

In the aforementioned non-technical introduction to game theory, Luce and Raiffa (1957) made it clear that vNM's idea of rehabilitating the representation of the individual decision problem through a numerical index called utility should not be misinterpreted. Although historically there had often been an unwise use of the numerical utility function—as a measure of something that economists should not really think they can measure—within a theory of decisions under risky conditions it has a precise meaning, as made explicit by the axiomatic approach used by vNM.

VNM's axioms guarantee that lotteries can be assigned numbers that reflect preferences, in the sense that one alternative has a higher utility than another alternative because it is preferred over the other, and not vice versa. Consequently, utility summarizes all the information about the preferences that generated the choices. By arguing in this way, every relevant aspect of the decision-maker's assessment of the probabilistic situation is taken into account, including the 'impure' pleasure associated with the game in itself, the gambling attitude that worried Smith, Jevons and Marshall. And this applies *a fortiori*, Luce and Raiffa (1957, p. 38) recalled, also to the other typical components on which economists had traditionally insisted. In particular, it was considered normal for economists examining the uncertain outcomes of an investment in financial assets to consider not only the average of prospective returns, but also their variability—summarized by the standard deviation or the variance—if not even higher-order variability indices. However, according to the new expected utility approach, there was no theoretical justification for this way of looking at portfolio choices.

Indeed, a typical decision outside the realm of choice under conditions of certainty is that of investment, not only investment in real capital goods, but also financial investment. And scholars studying the theory of financial markets were initially among the critics of expected utility theory. Friedman and Savage (1948, p. 281) recalled that Marschak himself had argued in one of his papers on the theory of financial assets, published before the *Theory of Games*, that it was "definitely unrealistic" to focus solely on mathematical expectation, as suggested by Bernoulli's "traditional calculus of 'moral probabilities'." But after vNM, this claim could no longer be justified.[19]

In fact, a coherent theory of portfolio choice, understood as the set of financial assets in which an individual's wealth is invested, did not yet exist in the 1940s. The

---

that is, seeing the other player's behavior in probabilistic terms. In mixed strategies, probabilities are introduced to prevent the other player from anticipating one's own behavior, not to assess with what probability the other player might choose a particular strategy.

[19] In his reconstruction of how the now largely accepted notion of the 'rational economic agent' came about, Giocoli (2003, p. 386) quotes Samuelson's recollection of his own reaction at the time of *Theory of Games*: "Why should you embrace the dogma of Expected Utility maximizing? Why not maximize the mode, the median or geometric mean of utility outcomes? ... From the highlighting in Marschak (1950) of an 'independence axiom' and from Savage's use of the words 'sure-thing-principle' I came to discern the appeal in behavior axioms that do lead to realization of the Expected Utility dogma." We will return to the so-called sure-thing principle in Chap. 5.

theoretical contribution that defined precisely how to deal with the issue of diversification, which economists had always evoked but never actually incorporated into their models, was due to the work of Markowitz. Harry Markowitz (1927–2023), a graduate of the University of Chicago and Marschak's student during his doctorate, noted that measures such as the simple maximization of an investor's objective function do not guarantee that the level of risk associated with the investment, represented by the variability of outcomes, is taken into account. The rule that investors should maximize the discounted value of future returns was inadequate because it did not admit any diversification among the assets to be held: it would merely be a matter of identifying the financial asset with the highest expected average return and investing everything in that asset. Of course, this rule is dubious in the real financial market, as it does not take into account the risk that concerns investors, who typically aim at a diversification of their holdings. Markowitz (1952) instead considered future return flows as random variables and examined the statistics representing them, focusing on mean and variance. He then defined an 'efficient portfolio' as one in which the maximum level of return (mean) is constrained to a given level of variability (variance), and demonstrated that the optimal choice is one in which investments are made in several financial assets simultaneously: the efficient portfolio respects the principle of diversification. Since then, modern portfolio theory has been built on this basis.

Markowitz developed these themes when at the Cowles Commission—the Chicago research center mentioned in the previous section—the discussion focused on game theory and, in particular, on the possible extension of vNM's expected utility maximization to choices under conditions of uncertainty. Yet, the theoretical result of vNM was in apparent contradiction with the Markowitz procedure, since expected utility incorporates every possible consideration inherent in the choices from which it is derived. Indeed, in the reconstruction of the development of his research that he would give in the lecture accepting the Nobel Prize, awarded to him in 1990, Markowitz recalled that while attending the meetings at the Cowles Commission in those years, he had to confront the problem of how to justify the employment of mean and variance. It was a question of checking the congruence between the theory of rational behavior that was emerging in those years and his theory of the efficient portfolio, which was designed more to describe the phenomenon of diversification than to provide a justification based on general principles.

Markowitz addressed this issue in a 1959 Cowles monograph entitled *Portfolio Selection. Efficient Diversification of Investments*. In the first part, on the basis of his 1954 doctoral dissertation, he presented a more complete version of his portfolio theory: by including previously neglected computational aspects, he showed how to broaden the scope of a theory that had been motivated and justified by the actual behavior of individuals in the financial markets, and was thus intended as a descriptive theory. But in the second part, he paused to test the compatibility of his theory with vNM's (and Savage's) principles of rational behavior, which he claimed to accept in full as a normative theory. The question was now approached from a different angle, Markowitz recalled (1991, p. 471). The question was that of assessing which function was a good approximation of an expected utility function:

> if an investor with a particular single period utility function acted only on the basis of expected return and variance, could the investor achieve almost maximum expected utility? Or, to put it another way, if you know the expected value and variance of a probability distribution of return on a portfolio can you guess fairly closely its expected utility?[20]

We saw in the previous paragraph that the basis of vNM's theory is that each future alternative can come true with some given probability. But if one moves away from the extremely simple case of two possible objects of choice, *A* and *C*, and their probabilities, α and $1 - \alpha$—the case we referred to in the previous sections—and assumes a number n of objects and associated probabilities, the matter becomes considerably more complicated. The problem becomes that of identifying an entire probability distribution. Just as in economics we are interested in a convenient—in the sense of mathematical tractability—representation of preferences, so we are interested in a convenient representation of the probability distribution. The assumptions that are typically made about probability distributions are justified by the possibility of representing them synthetically by means of parameters, precisely because, as the number of outcomes to be observed increases, the assignment of probabilities to outcomes generates increasingly complex probability distributions.

Markowitz (1959, Ch. 14) recognized that under two restrictive but widely accepted assumptions, it is possible to establish an equivalence between expected utility maximization and the efficient portfolio choice solution he had identified earlier. Assuming a quadratic utility function and a normal distribution (in the probabilistic sense) of the outcomes of the future returns of each asset, it is possible to show that portfolio theory is an application of the principle of expected utility maximization. For example, Markowitz showed that if there are two portfolios that are equal in mean and variance of expected returns, a third portfolio that is a combination of the two—a diversified portfolio—is certainly better than the first two, provided that the utility function is concave downward: risk aversion implies diversification.

Ultimately, although this issue is often overlooked, the established portfolio theory associated with Markowitz's name was accepted by economists precisely because its author was able to present it as a solution compatible with expected utility maximization. In the same years in which the theory of decisions under risky conditions was being developed—and, as we shall see in the next chapter, extended to conditions of uncertainty—a modern theory of portfolio choice emerged, which could be considered as one of its applications.[21]

[20] Markowitz (1991, p. 470) recalled that he began his analysis by assuming that an investor faces known probabilities. But "of course, none of us knows probability distributions of security returns" when investing in financial assets. Nevertheless, Markowitz continued, "I was convinced by ... one of my great teachers at the University of Chicago that a rational agent acting under uncertainty would act according to 'probability beliefs' where no objective probabilities are known; and these probability beliefs or 'subjective probabilities' combine exactly as do objective probabilities." The reference is to Savage and his theory of subjective expected utility, which we will discuss in the next chapter. But we have anticipated this consideration here because it is symptomatic of the fact that the idea that all uncertainty can be regarded as probabilistic risk, even in the financial markets, was gaining ground in the mid-1950s.

[21] It may be instructive to report here a well-known anecdote that Markowitz mentioned in the closing of in his Nobel lecture, concerning the defense of his doctoral thesis at the University of Chicago

4.6 Concluding Remarks

The advent of vNM's *Theory of Games* signaled a paradigm shift in the field of economics, marking its transition into a modern era characterized by a formal mathematical precision that sought to emulate the exact sciences. The axiomatic method—predicated on concepts and propositions whose meaning is in principle independent of the specific field of analysis—enabled the derivation of conclusions that could be substantiated through theorems, even in the domain of economics. Consequently, these conclusions could be accepted with the assurance that they had been derived accurately and were not mere conjectures.

This shift marked a departure from the prevailing traditional method of reasoning, which relied on the theorist's intuitions and subsequently developed through logical argumentation, or alternatively, involved the examination of a series of empirical cases without a structured theoretical framework as guide for the analysis. Indeed, Marschak (1946, p. 115) argued in support of vNM's methodology that by following the axiomatic method the "'intuitive' or 'heuristic' considerations generated in the minds of the authors by experience are formalized into concepts and propositions which, once stated, are detached from experience until the final conclusions are reached." Unlike traditional mathematical economics, which did not use such "meticulous" formalism,

> not only are conclusions obtained in this way more reliable, not only do they more audibly respond to any deliberate modification of the premises, but they are also richer and better ordered and their mutual relations are revealed with greater clarity than if one had studied one empirical case after another, encumbered with every accidental detail and exposed to the risk of misunderstandings and fallacies.

Even when guided by the use of mathematical tools, economists' analyses had not been immune to the danger of arriving at vague conclusions.

Since it was the basic result upon which game theory was developed, decision theory under risk was the first application of this novel way of thinking. However, the process could not be considered complete until the possibility of explaining uncertainty in similar terms was also assessed. Marschak himself, in his systematization of the expected utility theory mentioned above, still distinguished between decisions under conditions of "complete information," where the probabilities are known, and "incomplete information," where they are not (Marschak, 1950, p. 113). He acknowledged that his axiomatization of vNM's expected utility applied exclusively to the first scenario, while no systematic analysis was yet available for the second, identified as the one Knight had "supposedly" referred to with the term uncertainty. The next chapter is devoted to this fundamental passage.

---

in 1954. Among the members of the committee awarding the Ph.D. degree was Friedman, who at that time was still involved in Savage's research about how to extend the application of expected utility maximization (Friedman & Savage, 1952). At the end of the presentation, Friedman remarked that the degree to be awarded was in economics, and that a dissertation on portfolio choices did not qualify for a Ph.D. in economics. Markowitz (1991, p. 476) also revealed that he obviously obtained the title and that economists would later recognize, by awarding him the Nobel Prize, that portfolio theory could well be the subject matter of an economics dissertation.

References

Boumans, M., & Davis, J. B. (2010). *Economic methodology: Understanding economics as a science*. Palgrave Macmillan.

Dimand, R. W. (2020). The Cowles Commission and foundation for research in economics. In *The new Palgrave dictionary of economics online*. Palgrave Macmillan.

Ellsberg, D. (1954). Classic and current notions of "measurable utility." *Economic Journal, 64*, 528–556.

Ellsberg, D. (1956). Theory of the reluctant duelist. *American Economic Review, 46*, 909–923.

Fishburn, P. C., & Wakker, P. (1995). The invention of the independence condition for preferences. *Management Science, 41*, 1130–1144.

Friedman, M., & Savage, L. J. (1948). The utility analysis of choices involving risk. *Journal of Political Economy, 56*, 279–304.

Friedman, M., & Savage, L. J. (1952). The expected-utility hypothesis and the measurability of utility. *Journal of Political Economy, 60*, 463–474.

Giocoli, N. (2003). *Modeling rational agents: From interwar economics to early modern game theory*. Edward Elgar.

Giocoli, N. (2013). From Wald to Savage: *Homo economicus* becomes a Bayesian statistician. *Journal of the History of the Behavioral Sciences, 49*, 63–95.

Hicks, J. R. (1939). *Value and capital*. Oxford University Press.

Luce, D., & Raiffa, H. (1957). *Games and decisions*. Wiley.

Markowitz, H. (1952). Portfolio selection. *Journal of Finance, 7*, 77–91.

Markowitz, H. (1959). *Portfolio selection: Efficient diversification of investment*. Wiley.

Markowitz, H. (1991). Foundations of portfolio theory. *Journal of Finance, 46*, 469–477.

Marschak, J. (1946). Neumann's and Morgenstern's new approach to static economics. *Journal of Political Economy, 54*, 97–115.

Marschak, J. (1950). Rational behavior, uncertain prospects, and measurable utility. *Econometrica, 18*, 111–142.

Menger, K. ([1934] 1979). The role of uncertainty in economics. In Id. *Selected papers in logic and foundations, didactics, economics*. Reidel Publishing Company.

Mirowski, P. (2002). *Machine dreams. Economics becomes a cyborg science*. Cambridge University Press.

Morgenstern, O. ([1935] 1976). Perfect foresight and economic equilibrium. In A. Schotter (Ed.), *Selected economic writings of Oskar Morgenstern*. New York University Press.

Morgenstern, O. (1937). *The limits of economics*. Hodge.

Morgenstern, O. (1976). The collaboration between Oskar Morgenstern and John von Neumann on the theory of games. *Journal of Economic Literature, 14*, 805–816.

Moscati, I. (2016). Retrospectives. How economists came to accept expected utility theory. The case of Samuelson and Savage. *Journal of Economic Perspectives, 30*, 219–236.

Moscati, I. (2018). *Measuring utility: From the marginal revolution to behavioral economics*. Oxford University Press.

Moscati, I. (2023). *The history and methodology of expected utility*. Cambridge University Press.

Mosteller, F., & Nogee, P. (1951). An experimental measurement of utility. *Journal of Political Economy, 59*, 371–404.

Myerson, R. B. (1999). Nash equilibrium and the history of economic theory. *Journal of Economic Literature, 37*, 1067–1082.

Popper, K. ([1934] 1959). *The logic of scientific discovery*. Hutchinson.

Samuelson, P. A. (1938). A note on the pure theory of consumer's behavior. *Economica, 5*, 61–71.

von Neumann, J. ([1928] 1959). On the theory of games of strategy. In R. D. Luce, & A. W. Tucker (Eds.), *Contributions to the theory of games* (Vol. 4). Princeton University Press.

von Neumann, J. ([1937] 1945). A model of general economic equilibrium. *Review of Economic Studies, 13*, 1–9.

von Neumann, J., & Morgenstern O. ([1947] 2004). *Theory of games and economic behavior* (2nd ed.). Princeton University Press.

Wald, A. (1945). Statistical decision functions which minimize the maximum risk. *Annals of Mathematics, 46*, 265–280.

Chapter 5
Ramsey, De Finetti and Savage: Decisions Under Uncertainty

5.1 Introduction

In Chap. 3, we introduced Frank Knight's distinction between risk and uncertainty, together with the more elaborate, probabilistically grounded, version found in John Maynard Keynes's work. Chapter 4 was devoted to presenting the methodology employed by John von Neumann and Oskar Morgenstern (whom we will continue to refer to as vNM) in their investigation of the problem of rational choice in risky situations. Although vNM focused on a different topic, that of the strategic interactions between individuals, they proposed an axiomatic framework in which, under the assumption that the probabilities of outcome-relevant events are given, the structure of the choice problem was generalized. Following vNM, the scheme that had already been developed in economics for choices under conditions of certainty came also to be applied to choices involving risk.

Although it occupied only a few pages in a volume of over six hundred, for advocates of a mathematical approach to economic decisions, the derivation of expected utility was the most interesting aspect of the *Theory of Games*. Clearly, the possibility to treat all choice problems as if individual agents were utility maximizers, even when making decisions under risk, was a particularly convenient simplification. Moreover, the immediate appreciation of this result was indicative of a general tendency toward formal rigor that characterized much of the development of economic theory in the immediate post-World War II period. Economics had emerged from World War II with the aura of a rigorous science employing general methods for solving practical problems. The young mathematical economists of this period were eager to demonstrate that further use of mathematical tools would enhance the explanatory power of economic theory.

For instance, Duncan Luce and Howard Raiffa—who were part of this group of mathematical economists—emphasized in their aforementioned volume on games and decisions that the employment of new mathematical tools was now widespread. There had already been important developments for problems under conditions of

C. Zappia, *Uncertainty in Economics*, Springer Studies in the History of Economic Thought, https://doi.org/10.1007/978-3-031-91221-4_5

certainty, starting with the technique of linear programming as a method for solving resource allocation problems. During the war, the choice of routes that minimized the distances traveled by convoys of a fleet transporting military vehicles between predetermined harbors had been guided by techniques suggested by economists. Luce and Raiffa (1957, pp. 17–18) argued that this was an example of a general problem, that of trying to determine which action, among those meeting certain criteria of feasibility, was the least costly. They emphasized that the general problem had an immediate solution if one employed the theory of linear spaces, whereas the traditional use of differential calculus offered no assistance in such scenario. Moreover, the solution was obtained using the same technique developed for an entirely different problem, that which von Neumann had developed for a particular type of game, the two-person zero-sum game. Notably, any linear programming problem could be associated with a strategic problem, with the solution of one problem being formally equivalent to the solution of the other. The new mathematical tools were general in the sense that they were applicable to various contexts.

This was the period when economics really became a mathematical science. According to Tjalling Koopmans—who succeeded Jacob Marschak in the late 1940s as director of the Cowles Commission for Research in Economics, which we mentioned earlier as the leading research center in mathematical economics—the very allocation of resources within an economic system could be thought of as a huge linear programming problem. It was from the study of linear programming—also called activity analysis—that a stream of research developed that culminated in the demonstration of the existence of a general economic equilibrium. That it was possible to identify a set of prices at which the markets for all goods are in equilibrium at the same time had been hypothesized by Léon Walras at the end of the nineteenth century, but never proved mathematically until the work of Kenneth Arrow and Gérard Debreu, both researchers associated with the Cowles Commission at the time (Weintraub, 2002).[1]

In this manner, economics endeavored to establish itself as an exact science, akin to the natural sciences: a science capable of offering general theoretical frameworks that provide solutions to complex practical problems. As previously stated, vNM demonstrated that risk could be treated in a manner analogous to certainty. This entailed assuming that economic agents, in adhering to a set of axioms defining rationality, would act as if they were maximizing a utility function. However, a further fundamental issue remained to be addressed, one which Keynes had already emphasized in his *Treatise on Probability* but which was only hinted at in vNM's *Theory of Games*: for a comprehensive theory of decisions, the really challenging cases are those in which the probabilities are not objectively given.

---

[1] Düppe and Weintraub (2014, p. 454) argue that the 1949 conference on "Activity analysis of Production and Allocation," organized by Koopmans to extend the theory of linear programming developed in wartime research to the economic theory of production, can be considered "the 'coming out party' of the community that would transform the practices of academic economists for decades to come."

Crucial economic decisions, such as whether or not to undertake a particular investment, whether or not to accept a job offer, and others of similar import, involve probabilities that are inherently subjective. The probabilities associated to the events that condition the expected outcomes are "personal" to the individual making the decision, as Leonard Savage ([1954] 1972, p. 30], the main contributor to this topic, would call them. The question, therefore, is how to assign probabilities to specific single events whose occurrence cannot be meaningfully associated to past observations—as opposed to when there are empirically detectable frequencies—and which cannot be framed in contexts that satisfy conditions of symmetry—as is the case with games of chance. According to Knight, as we have seen, this was not possible. However, for mathematical economists after World War II, the question could be solved through the adoption of an approach similar to that employed by vNM, namely, on an axiomatic basis.

Keynes had taken a somewhat intermediate position between Knight's outright rejection and a precise numerical approach of the kind that vNM would later justify on the basis of axioms, but which was already implicit in the expected utility associated with utilitarian ethics after Daniel Bernoulli, although disputed by the marginalist economists of the late nineteenth century. Keynes had attempted to adopt a logical approach to the problem of how to deal with subjective probabilities, while at the same time drawing inspiration from ordinary reasoning. In ordinary reasoning, it is typical to formulate beliefs in such a way that one event is considered more or less probable than another, without being able to say how much more or less probable it is—in the sense of assigning to it a precise, numerical probability. However, after characterizing the problem, Keynes had stopped at a critique of the decision criterion of expected utility maximization—later revalued by vNM for conditions of risk—without concretely identifying an alternative decision criterion for choices under uncertainty.

This chapter illustrates the emergence, in the early 1950s, of a solution to the problem of choice under uncertainty. This solution was immediately endorsed by decision theorists and subsequently accepted by the majority of economists, becoming a common feature of advanced microeconomics textbooks. However, as mentioned in Chap. 3, this solution had already been anticipated in a fundamental critique of Keynes's *Treatise* by the young English mathematician Frank Ramsey, and concurrently, but independently, developed by the Italian mathematical statistician Bruno de Finetti.

Taking a step back in time from vNM's contribution, this chapter begins with an examination of Ramsey's argument and then presents de Finetti's approach, both of which were formulated in the late 1920s. Although neglected at the time—with vNM developing their argument without taking them into account—these analyses remain highly relevant, as they provide the analytical and methodological foundations for subsequent developments. After two separate sections devoted to the contributions of Ramsey and de Finetti, Sect. 4 of this chapter is devoted to the definitive systematization of the subjectivist approach, which was presented by the American mathematical statistician Savage in the early 1950s, following in the footsteps of Ramsey and de Finetti.

5.2 Ramsey and Subjective Probability

In the Cambridge of the 1920s, which, as noted in Chap. 3, was characterized by circles of intellectuals from various disciplines, the figure of Keynes was central. Although he had not succeeded Alfred Marshall in the chair of Political Economy, which was held by Arthur Cecil Pigou from 1908 to 1943, Keynes's role was widely recognized and associated not only with the courses he taught but also with the weekly evening seminars he hosted in his rooms at King's College. The seminars were open to the most promising economics students, but the young mathematician Frank Plumpton Ramsey (1903–1930) was also admitted.

The son of a mathematics professor and an outstanding student in his high school years, Ramsey was valued from his first meeting with Keynes in 1921 for his ability to apply mathematical logic to economics. He epitomized the 'modern' mathematician, no longer focused on solving individual real-world problems, but able to bring logical rigor and consistency to any discipline with which he was confronted. Among economists, Ramsey is best remembered for two papers published in the *Economic Journal*, a journal to which he also collaborated at Keynes's request. In a first essay, Ramsey (1927) laid the foundations of the modern theory of optimal taxation, while in a second he introduced a growth model for determining the optimal rate of saving in the long run (Ramsey, 1928). Both are still regarded as some of the most brilliant works in mathematical economics, originating two distinctive fields of study. However, Ramsey also excelled in epistemology and philosophical studies. Notably, he translated from German the highly influential *Tractatus Logico-Philosophicus* by the Austrian philosopher Ludwig Wittgenstein, a major representative of logical positivism.

In 1926, Ramsey, who had already become a lecturer and director of mathematical studies at King's College, presented an essay on probability that began with a frontal attack on Keynes's *Treatise.* The essay, entitled "Truth and Probability," remained in draft form due to Ramsey's many other interests and his sudden death in 1930. Published posthumously in a volume collecting his other mathematical and philosophical essays, Ramsey's ([1931] 1964) essay on probability remained little known outside Cambridge even afterward, at least until its recognition as a founding text of subjective probability theory made by Savage in his 1954 volume. In his essay, Ramsey not only reiterated the criticisms he had already made of Keynes's approach in a 1922 review of the *Treatise*, but also presented the elements of a possible alternative theory, that of an epistemic approach to probability, but a purely subjective one rather than a logically determined one as in Keynes.

Ramsey was dissatisfied both with the introspective approach typical of the English utilitarians and with Keynes's attempt to justify the probabilistic relationship between a premise and a conclusion on logical grounds, an attempt which he felt was still too much based on intuition. Moreover, he did not believe that Keynes had really been able to ground his notion of probability. He agreed with Keynes that the justification of probability given by frequentists was inadequate in many relevant cases, but Keynes had assumed that rational individuals have the ability to perceive the degree to which a premise justifies a conclusion, something that seemed to Ramsey to be devoid of empirical content. Thus, he did not share with Keynes the idea that logic allows individuals to ascribe a character of rationality to opinions. According to Ramsey, there are no 'rational' degrees of belief *à la* Keynes, but only degrees of belief peculiar to the individual who formulates them, much like individual preferences. Of these beliefs, one can only say whether they are consistent or not, but not whether they are rationally justified.

Ramsey's argument was based on the idea that subjective probabilities could be justified only in terms of the actions to which they gave rise. It is from the choices made in uncertain situations that the probability assigned to an event can be derived, according to an approach inspired by pragmatist philosophy—from the Greek *praxis* for practical activity, as distinguished from speculative activity.[2] According to Ramsey ([1931] 1964, p. 71), a degree of belief is represented by "the extent to which we are prepared to act on it," and can therefore be measured by reference to an individual's behavior, whether in real or hypothetical situations. To construct a mathematical theory of beliefs, he proposed to employ a "general psychological theory," the theory that "we act in the way we think most likely to realize the objects of our desires, so that a person's actions are completely determined by his desires and opinions" (Ramsey, [1931] 1964, p. 74). Desires refer to preferences (utilities) about the consequences of actions, while opinions refer to degrees of belief (probabilities) about the events that condition the outcome of actions. Consequently, in situations of uncertainty, individuals can be assumed to act by maximizing the mathematical expectation of the consequences of their actions.[3]

Ramsey argued that it is well known that the probabilities of events can be inferred from the activity of bettors, who, by agreeing to take part in a bet on a given event, are in fact revealing the probability they assign to the occurrence of that event. As a matter of fact, every human action is a kind of gamble:

> whenever we go to the station we are betting that a train will really run, and if we had not a sufficient degree of belief in this we should decline the bet and stay at home. The options God gives us are always conditional on our guessing whether a certain proposition is true. (Ramsey, [1931] 1964, p. 81)

More specifically, Ramsey ([1931] 1964, pp. 75–76) suggested the following reasoning. Suppose you have come to a fork in the road and you do not know which route to take to reach a certain place, but you decide to go in the direction you think is the most correct, while remaining on the alert in case there is a resident whom

---

[2] Ramsey's reliance on pragmatism was derived from the American philosopher mathematician Charles Sanders Peirce, for whom the essence of pragmatism was that "the meaning of a sentence is to be defined by reference to the actions to which asserting it would lead, or, more vaguely still, by its possible causes and effects." This quote is taken up by Misak (2020, p. 261), who emphasizes the crucial role of pragmatism for Ramsey's approach. As we will see, this adherence aligns with the operationalism explored by de Finetti and subsequent decision theorists.

[3] Ramsey ([1931] 1964, p. 74) acknowledged that this theory was "universally discarded" at the time, with an implicit reference to the Marshallian tradition we highlighted in Chap. 2. He argued, however, that he believed it to be "fairly close to the truth in the sort of case with which we are most concerned."

you can ask for information. If you see that there is a resident in a field at a certain distance away who certainly knows the place, you might make a detour to ask the resident what the correct route actually is. Ramsey proposed to consider the distance one is willing to travel to get the information as a measure of confidence that the route taken is correct.

If you call r the utility of arriving at the right destination and w the utility of arriving at the wrong destination, with $r > w$, and if $f(x)$ is the (dis)utility to you of traveling x meters away from the direction you took, then you can determine the distance you are willing to travel. In fact, if you do not ask for information, with some unknown probability p you will obtain a result r, and with $(1 - p)$ a result w, that is: $pr + (1 - p)w$. If you ask for information, as a function of distance you will certainly obtain $r - f(x)$. Deciding to ask therefore means assuming that $r - f(x) > pr + (1 - p)w$, from which you deduce that the maximum distance you are willing to travel to ask for information is that x^* for which $r - f(x^*) = pr + (1 - p)w$, i.e., $f(x^*) = (r - w)(1 - p)$. But if you rewrite this relationship as $p = 1 - f(x^*)/(r - w)$, you would see what maximizing expected utility entails: given the utilities r and w, the distance you are willing to travel to ask for information reveals the probability that you assign to the event that the road you took is the correct one.

Of course, no theory can be built on an example, however illuminating it may be. Ramsey was interested in identifying the general properties that a function measuring degrees of belief must satisfy in order to be identifiable as a probability. We noted in the previous chapter that vNM proposed a method whereby choices about lotteries, whose outcomes could be obtained with objectively known probabilities, yielded a measure of the utility implicitly attributed to the objects of choice. We also noted that von Neumann was surprised that such a simple method of deriving a utility function from choices over lotteries had not already been proposed by others. In fact, the procedure that Ramsey imagined in his 1926 paper was essentially the same as the one that would be accepted by economists in the late 1940s, after being presented in vNM's *Theory of Games*. Moreover, as we shall see, de Finetti was also thinking along these lines.

But whereas vNM assumed given probabilities in order to derive utilities from choices, so that their reasoning applied only to situations of risk, Ramsey's idea was that, given utilities, probabilities could be derived from the choices, so that his reasoning also applied to situations of uncertainty. The novelty of Ramsey's approach, which would remain largely unexplored until Savage's work in the early 1950s, was therefore the identification of a way of measuring degrees of belief that did not presuppose any reference to objectively given probabilities, indeed, "a purely psychological method of measuring belief" (Ramsey, [1931] 1964, p. 69).

It should be noted that Ramsey's methodology simultaneously solved the problems of how to ground the notion of subjective probability and how to measure it. For frequentists, probability is defined by the frequency with which an event occurs, represented by the ratio of the number of times in which the event occurs to the total number of observed cases; the frequency is a precise number between zero (the event never occurs) and 1 (the event always occurs). For Keynes, probability is a logically derived degree of belief on the occurrence of an event, and is not always a precise

Fig. 5.1 Ramsey's derivation of subjective probabilities

	WIN	LOSE
BET	u(x)	u(0)
DO NOT BET	u(y)	u(y)

number. More often it is an interval, with the indeterminacy being due to the lack of objective data assumed by the frequentists. According to Ramsey, instead, one can only define individuals' probability by referring to the actions that result from their beliefs. And while explaining where probabilities come from, he also provided a method for calculating them precisely. Thus, Ramsey's new approach amounted to a critique of both the logical foundations of Keynes's probability and his idea that probabilities are not precisely measurable in all uncertain domains. Indeed, Ramsey argued ([1931] 1964, p. 70), a degree of belief has "no precise meaning unless we specify more exactly how it is to be measured."

Recalling that it was traditionally accepted that one "way of measuring a person's belief is to propose a bet, and see what are the lowest odds which he will accept," Ramsey ([1931] 1964, p. 73) proceeded by comparing two situations x and y, the first of which is uncertain and the second of which is certain.[4] If the individual derives an uncertain utility $u(x)$ from x and a certain utility $u(y)$ from y, it is possible to evaluate his degree of belief in the event that will win him x as the ratio $u(y)/u(x)$. Winning x will occur with some unknown probability if an event occurs—e.g., a particular horse on which the individual has bet is the winner of a race—while y has a known probability of 1—which would identify the sure amount of money one can keep by not betting on the race.

It may be worth summarizing: if the relevant utilities are assumed to be known, then the observed choices can be used to infer the subjective probability values, that is, to measure probability. This can also be seen by using the same matrix representation that we used in Chap. 2 to illustrate Bernoulli's argument (see Fig. 5.1). The individual is asked to indicate which amount of money y gives him a utility value $u(y)$ such that he is indifferent between betting and not betting. And if he indicates a given y, by application of the "general psychological law" of maximizing expected utility this would imply that $pu(x) + (1 - p)u(0) = u(y)$, i.e., $p = \mathrm{u}(y)/\mathrm{u}(x)$.

For instance, if the individual is indifferent between gaining, say, 10 units of utility by participating in the bet and having, say, 4 units of utility by not betting, then he reveals a subjective probability of the uncertain event of $4/10 = 0.4$. Thus, if the bettor offered him a ticket for the bet that costs even slightly more than 4 utility units, the individual would not accept it; but if he was offered the ticket at a price slightly less than 4 utility units, he would accept it. A ticket bought at a utility of 4 corresponds to the individual's reservation price, or, as Ramsey would assert, 4 utility

[4] In English betting jargon an event is quoted in the form of, for example, 2:3 (2 against 3) to indicate that by betting 2 on the occurrence of the event one can win 3. By normalization, these odds correspond to a probability of the event occurring equal to $2/(2 + 3)$, or $2/5$.

units is the price that reveals that the individual is indifferent between the uncertain event and the certainty of his initial endowment.[5]

Ramsey did not refer to Thomas Bayes, but his definition of subjective probability coincides with that used by Bayes in his seminal essay on the "doctrine of chances," an essay that is universally famous because of the introduction of an argument about conditional probabilities, which implied the so-called Bayes Theorem (Bayes, [1763] 1958). In fact, Bayes's definition of a priori probability had not attracted much attention in general, remaining embedded in the classical definition of probability with which Pierre Simon de Laplace had associated it. Neither Ramsey nor, as we shall see, de Finetti had explicitly taken it up again, although what they proposed turned out to be essentially a reformulation of it, on axiomatic grounds that Bayes had of course not imagined.[6]

Ramsey was therefore reviving a tradition—soon to be identified as 'Bayesian'—in which, if this way of deriving probabilities is indeed valid, while on the one hand a subjective probability can be assigned to any kind of event, on the other hand there is no reason why the probabilities assigned to a given event by one individual should be the same as those assigned to it by another individual. According to Keynes, two individuals endowed with logical capacities must necessarily derive the same degree of belief in a given proposition from the same evidence, and the probability is unique in that it is rationally derived. According to Ramsey, on the contrary, this need not be the case: from this point of view, Ramsey's approach can be said to be psycho-logical, rather than logical as in Keynes. Degrees of belief are not rational, but subjective in the sense of being personal, and can therefore vary from individual to individual.[7]

---

[5] Ramsey was aware, of course, that he could not reason directly in terms of money, because the tradition of marginalist thought dictated that the marginal utility of money was always decreasing. He therefore proceeded to an axiomatization of utility in terms of what he called an "ethically neutral event." Having defined as neutral an event about which the individual is indifferent as to whether to bet for or against—therefore assigned a subjective probability of ½—Ramsey proposed to derive utilities for monetary outcomes by observing choices between gambles conditioned on the realization of the neutral event. For simplicity, we have omitted this step and used the idea that gains are directly represented in units of utility, but see Newman (1987) for a clear summary of the procedure followed by Ramsey. Obviously, a more direct way of arguing could be to use monetary amounts, with the proviso that the gains be taken as very small amounts of money relative to the initial monetary wealth of the individuals, so as to make the value of marginal utility unimportant. As we shall see in the next section, this is what de Finetti assumed in his investigation.

[6] In Bayes's ([1763] 1958, p. 298) words: "The probability of any event is the ratio between the value at which an expectation depending on the happening of the event ought to be computed, and the value of the thing expected upon its happening."

[7] Such a reinterpretation of Bayes, which allows not only the utility of outcomes but also their probability to vary subjectively between two people evaluating the same gamble, was also leveled against Keynes by the French mathematician Émile Borel. In a review of the *Treatise* originally published in French, Borel ([1924] 1964, pp. 50–51) argued: "It is frequently said of two bettors that one is a thief and the other is an imbecile; this is true in certain cases, where one of the two bettors is better informed than the other, and knows it; but it can also happen that two men in good faith in complex situations where they possess exactly the same elements of information will arrive at different conclusions on the probabilities of an event and that betting together, each one figures, to use a picturesque but exaggerated expression, that he is the thief and the other the imbecile."

Nevertheless, subjective does not mean arbitrary: for degrees of belief to have the analytical value of a probability, they must be consistent. Indeed, in Ramsey's view, the accurate identification of probabilities is linked to the formulation of what he called a "logic of consistency."

The issue of the consistency of subjective probabilities is an important one, and will be addressed several times later in this history, so it may be useful to introduce it with an example. Suppose a person goes to a race in which three horses participate and has paid 0.4 utility units (but would not have paid a slightly higher price) for a ticket that guarantees her 1 utility unit if a particular horse wins, say Horse A. We have just seen that for Ramsey, a person's belief in the occurrence of an event can be deduced from the lowest price that induces her to accept the bet. So the person who is willing to pay this price reveals that she has a subjective probability $p(\textit{Horse A wins}) = 0.4$. But then if the same person is offered to bet on the event that one of the other two horses wins, she would have to be willing to pay at most 0.6 to win 1 if one of the other two horses wins, thus revealing a probability $p(\textit{Horse A loses}) = 0.6$. Otherwise, if she revealed, say, $p(\textit{Horse B wins}) = 0.35$ and $p(\textit{Horse C wins}) = 0.35$—i.e., $p(\textit{Horse A loses}) = 0.7$—one could structure a set of bets for which she would be willing to pay $0.4 + 0.7$ to win at most 1, thereby incurring sure loss. This would be a violation of the additivity property of probabilities recalled in Chap. 2.

In fact, in what can be considered the first clear statement characterizing what came to be known as the Bayesian approach to decision theory, Ramsey ([1931] 1964, p. 80) observed:

> If anyone's mental condition violated these laws, his choice would depend on the precise form in which the options were offered him, which would be absurd. He could have a book made against him by a cunning better and would then stand to lose in any event.

Since a betting structure that implies a sure loss for one of the parties is traditionally called a Dutch book—presumably in reference to the ability to profit from arbitrage that English merchants attributed, with disdain, to Dutch merchants—the consistency condition imposed by Ramsey is usually referred to as 'the Dutch book argument' (Hájek, 2009).[8]

In sum, Ramsey presented an idea of probability that was built into a logic of consistency, where the probability measure was derived from degrees of belief as revealed by actions. That degrees of belief were coherent, so that they could be represented by a function with the standard properties of probability, was the result guaranteed by the Dutch book argument. And this enabled the employment of the criterion of maximizing expected utility even when objective probabilities were not given. In

[8] We put ourselves in the position of the person who has to indicate the price at which she is willing to take part in the bet proposed to her. Of course, it is also possible to reason the other way round, from the point of view of the person proposing the bet, which makes the argument even more straightforward. Bookmakers who offer to bet on Horse A to win at a price of 0.4 and at the same time to bet on Horse B to win at, say, 0.25 and Horse C at 0.25, thus to bet on Horse A to lose at a price of 0.5, would be putting themselves in a position to lose in any case. In fact, a bettor could buy all the tickets at 0.9 and be sure of getting 1, which would mean a sure win (and, in the long run, bankruptcy for book holders).

fact, it was possible to reason about the individual's choices in this way because a method had been identified to make sense of subjective utilities *and* probabilities, that is, to measure them precisely.

That Ramsey was presenting Cambridge scholars with a method of identifying degrees of belief with probabilities so innovative that it would later lead to a hegemonic interpretation of probability theory could not have been entirely clear to Keynes. It was only after the publication of the work of de Finetti in French in 1937 that this view began to gain the status of a distinctive subjective approach to probability.[9] However, Keynes soon realized the value of the criticism he had received and, in remembering the young collaborator after his premature death, he recognized the importance of Ramsey's approach as a cogent alternative to his own.

In his obituary for the *Economic Journal* in 1930, Keynes recalled that Ramsey had contributed to several fields of research, but that the sophisticated mathematical tools used by Ramsey were particularly significant in presenting the potential of a logical-mathematical approach to economics. Showing the highest respect for his young fellow, Keynes argued that the two papers Ramsey had published in the *Economic Journal* indicated to economists a difficult path to follow, but one that could provide relevant results with immediacy and precision. More relevant to our story, Keynes also commented on Ramsey's contribution to the philosophy of probability in a short essay of the following year reviewing the posthumous publication, in 1931, of Ramsey's philosophical essays—a comment later published along with a series of biographical essays (Keynes, [1933] 1972). In his review, Keynes agreed that the method of calculation proposed by Ramsey ensured that, in principle, a numerical probability could be derived in any situation, and that this was a substantial step forward in the theory of probability. Ramsey's proposed new logic of the consistency of beliefs, understood as subjective—albeit not logical—probabilities seemed convincing: "So far I yield to Ramsey," Keynes ([1933] 1972, p. 339) wrote, "I think he is right." But one problem that Ramsey could not solve, Keynes continued, was that it was not clear what the empirically relevant 'true' probability might be, since subjective valuations might differ from individual to individual, even when based on the same evidence.

Ramsey had, indeed, wrestled with this question. His essay on probability ended with a section on the "logic of truth," which began with the observation that "to ensure that our beliefs are not self-contradictory"—the result that he had obtained in dealing with "logic as the science of rational thought"—was not enough: "we want our beliefs to be consistent not merely with one another but also with the facts" (Ramsey, [1931] 1964, p. 87). Some beliefs might be considered more relevant than others, Ramsey argued, because they more often correspond to what actually happens in reality. Thus, in addition to "formal logic"—that of consistency—there was a "human logic"—that of truth—through which probabilities could be evaluated

[9] Still in 1939, Ernest Nagel's *Principles of the Theory of Probability*, probably the first handbook to point out that a peculiar subjective view had begun to emerge, merely mentioned Ramsey and de Finetti as modern adherents of a classical Laplacian view, and considered their approach to be based on a "dubious psychological theory" (Nagel, 1939, pp. 18, 46).

according to the success associated with the actions they had induced. Here, too, Ramsey's pragmatism became apparent: some probabilities could be said to be more "reasonable" than others because we know from experience that they are more useful than others, and so we use them out of "habit." Ramsey seemed to be referring to an empirical substratum similar to that appealed to by frequentists, but the issue was barely touched upon in the essay, which, as already noted, was not published by Ramsey in his lifetime.[10]

In his review, Keynes objected to this point. He argued that unless this aspect of probability theory was convincingly addressed, Ramsey's approach could not really answer the questions that preoccupied him. In a sense, Keynes ([1933] 1972, p. 339) concluded in his assessment of his young collaborator's research, Ramsey had indeed given content to the idea of probabilistic precision and he "may have been pointing the way to the next field of study when formal logic has been put in good order and its highly limited scope properly defined." But he had done so by stretching the possibilities of the application of a subjective theory of probability to the point of losing all reference to what can be rationally deduced about reality. In the end, Ramsey could only restore the link with reality by abandoning to some extent the subjective approach, with what could only have appeared to Keynes as an appeal to frequencies. Thus, Keynes concluded his review with a distinctly dismissive tone: it is not by asserting that choosing one probability over another is a "useful mental habit" that one gets to the "bottom of the principle of induction," which, as we know from Chap. 3, for Keynes consisted in deriving probabilities from the available evidence.[11]

Keynes's brief remarks on Ramsey's probability theory were terse, and have therefore been the subject of extensive and conflicting analysis, also because they have been considered crucial to understanding the extent to which Keynes remained committed to his early philosophical position when writing the *General Theory* (Gerrard, 1992). According to some interpreters, Keynes did indeed yield to Ramsey, abandoning both the logical approach and the idea that there might exist non-numerical probabilities. It can be argued, however, that for Keynes the solution could be found in a different, generalized logic of consistency. Although Keynes may have admitted in his comments on Ramsey that he no longer adhered to his epistemic-but-objective version of probability theory, he thought it possible to preserve the formal apparatus of his theory by adopting a theory of probability with beliefs derived from actions, but not as definite and consistent as Ramsey had assumed (Runde, 1994; Kyburg, 2003).

Indeed, had Keynes conceded in 1931 that Ramsey's criticism of the *Treatise* was correct, it would have been impractical to construct a line of continuity between the probability of the 1921 *Treatise* and the uncertainty of the 1936 *General Theory*,

[10] The potential shift from rationality to reasonableness in Ramsey's subsequent work, as evidenced by a series of scattered notes he wrote prior to his sudden death, has been highlighted in Gaspard (2014) and Misak (2020).

[11] On the basis of Ramsey's further notes on the issue of chance, Galavotti (2011) posits that Ramsey sought to develop an original methodology for interpreting objective probability, particularly in the context of chance and probability in physics, in terms of degrees of belief, an attempt that Keynes might have overlooked.

as we did in Chap. 3. Notably, the critical tone of Keynes's concluding remarks implies a persisting conviction that degrees of belief that cannot be quantitatively expressed should be taken into account, particularly in contexts involving the 'very uncertain' scenarios alluded to in the epigraph preceding the introduction to this volume and examined in Chap. 3. The message of the chapter of the *Treatise* dealing with the application of probability to conduct—that the maximization of expected utility is an inadequate criterion under conditions of uncertainty—does not appear to be questioned by Keynes, even in light of Ramsey's criticism.[12]

However, irrespective of Keynes's real judgment about Ramsey's subjective probability, advancements in this domain were on the verge of being acknowledged by the new mathematical economists working on decision-making after World War II. This recognition was further facilitated by the fact that the same issues as Ramsey's had been addressed by another scholar, the Italian mathematical statistician Bruno de Finetti.

5.3 The Logic of the Probable According to De Finetti

We have just seen that the method of choice for risk conditions, axiomatized by vNM in 1947, had in fact already been proposed for the more general setting of uncertainty by Ramsey in the late 1920s. Ramsey opened up the study of probabilities that were not purely logical, as Keynes had assumed, but determined by a person's inclination to act, hence psycho-logical; and yet not even purely psycho-logical, since they were in any case probabilities that represented a peculiar kind of rational behavior, in the sense of obeying a logic of consistency. But Ramsey's contribution was not largely recognized until after vNM revived the use of expected utility.

In a series of papers published in Italian and therefore initially little known, Bruno de Finetti (1906–1985) also devoted himself to the study of probabilities suitable for use in situations of uncertainty. De Finetti deserves considerable attention both for the formal aspects he introduced in the discussion of probabilities and for the context in which he presented them, making him a proponent of a radically subjective philosophy of probability. His role as a founder of the approach was recognized before that of Ramsey, because he was active in the international arena from the late 1930s.

Born in Innsbruck, Austria, to Italian parents, de Finetti graduated in applied mathematics from the University of Milan in 1927. Although he began publishing on statistical issues soon after graduation, he could not embark on an academic career because his unmarried status made him ineligible under the Fascist regime. It was not until the late 1940s, after working for many years as an actuary and statistician

[12] As already noted in Chap. 3, the exchange with Hugh Townshend following the publication of the *General Theory* appears to substantiate this claim. In this exchange, Keynes ([1937–38] 1979, p. 289) reiterated to Townshend, who seemed to be leaning toward a criterion for decision under uncertainty analogous to the certainty equivalent, and thus taking into account both utilities and probabilities in the style of Ramsey, that if subjective probabilities were to be used, it should not be forgotten that "on my theory of probability, the probabilities themselves [...] are not numerical."

for the major insurance company Assicurazioni Generali in Trieste, that de Finetti became a lecturer in financial mathematics at the University of Trieste and then a full professor in calculus of probability at La Sapienza University in Rome. Although a French version of his early works published in 1937 in the *Annales de l'Institut Henri Poincaré* had a wide resonance among probability scholars, his international fame among economists, like that of Ramsey, owes much to Savage's definitive systematization of the subjectivist approach in the 1950s. Partly at Savage's urging, since then de Finetti devoted himself to the defense of the subjective approach, trying to overcome the hostility of the statisticians of the time, who were much more bound to the frequentist approach than were decision theorists and mathematical economists.

Theoretically, de Finetti's work is largely analogous to Ramsey's, but de Finetti placed more emphasis on the philosophical assumptions behind the subjective approach to probability, which he later developed in depth over the years, something that Ramsey was unable to do because of his early death. In addition, de Finetti worked extensively on the statistical application of subjective probability theory. Indeed, it is in this area that his contribution is fundamental: he showed in what sense, even within a subjective approach, it is possible to employ the sample observations typical of frequentist statistics.[13] For our purposes, however, it is interesting to revisit de Finetti's approach because he also addressed the implications of subjective probability for insurance decisions under uncertainty.

Like Ramsey, de Finetti worked under the influence of pragmatism, which he appreciated by reading the contributions of Italian mathematicians such as Giovanni Valiati and Federigo Enriques, who were inspired by Peirce as much as Ramsey. According to de Finetti, science must be useful to enable individuals to operate in the real world, but it does not consist of 'absolute' truths about the future occurrence of a fact; rather, it consists of knowledge that can be used to formulate predictions about the occurrence of that fact. Both experience and scientific elaboration contribute to the formation of a prediction, but the tool that allows predictions to be made is probability. In this sense, probability is the cornerstone of knowledge, both in science and in everyday life.

De Finetti's philosophical thought was therefore characterized by a rejection of determinism and transcendental laws, which he considered to be devoid of empirical

[13] The interpretation of a sequence of independent events differs between the frequentist and the subjectivist statisticians. For the subjectivist, by application of the Bayes Theorem, each new piece of information implies a revision of the initial probability, even that of the realization of a single draw from an urn whose composition is unknown. The sequence in which independent events are realized is therefore relevant to the subjectivist, but not to the frequentist. But De Finetti ([1937] 1964, pp. 118–120) argued that in reality there are a large number of situations in which independent events are actually "exchangeable events," in the sense that the order in which they occur is not really relevant. For the class of exchangeable events, De Finetti showed that as observations increase, subjective probabilities and objective probabilities, which apparently should differ, actually converge to the same value. In particular, de Finetti showed that the limits theorem of classical statistical inference hold just as well for sequence of exchangeable events as they do for independent, equiprobable events. On de Finetti's so-called exchangeability theorem, see Kyburg and Smokler (1964, pp. 12–13) and Galavotti (2001, pp. 199–203).

content, an example of an empty metaphysical philosophy, even in great thinkers such as Emanuel Kant or Karl Marx. He advocated an empiricist approach in which there are no cause-and-effect relationships proper to a rationalist view of science, but only probabilistic knowledge generated by a science that deduces the probable from necessarily partial observations (Galavotti, 1989). In this sense, he considered it meaningless to regard probability as an objective property, that is, as belonging to the observed phenomena and independent of the individual who evaluates them. Probability exists only insofar as a person uses it to make predictions, and is formulated on the basis of that person's partial knowledge of the course of events. Without the perspective of the individual, there is no probability, hence the famous statement de Finetti (1975) printed in capital letters in the preface to his book *Theory of Probability* to distinguish himself from the frequentists: "Probability does not exist."

Of course, de Finetti admitted that in order to assess the probability of an event, it is necessary to take into account all the available data, including possibly the symmetry conditions and frequencies emphasized by classical and frequentist statisticians. But any objective data is meaningful only when interpreted by the individual subject, and in any case there is an ineradicable subjective component due to the circumstance that each event is in fact a singular, unrepeatable fact. This is all the more true when dealing with situations of scarce and partial information, as is typical in the field of economics, to which de Finetti dedicated a great deal of space in his mature work, especially with regard to the theory of insurance contracts. In insurance theory, it is usually assumed that, when the insurer offers a contract to a particular individual who wishes to insure himself against the possible occurrence of an event that has a negative impact on the results of his activity or on his wealth, the insurer can refer to a class of similar cases. In reality, however, the insurer has to determine the probability of the event occurring in relation to the case of concern of each individual. As we shall see later, de Finetti worked this out as an application of his theory of probability.

Building on his early works in Italian, de Finetti outlined the subjective approach to probability in his seminal essay in French, later translated in English as "Foresight: its logical laws, its subjective sources."[14] Inspired by Henry Poincaré's dictum that "predictions can only be probable" and thus "all sciences would be but unconscious applications of the calculus of probability," de Finetti ([1937] 1964, pp. 99–100) aimed to show "how the logical laws of the theory of probability can be rigorously established within the subjectivist point of view." According to de Finetti, the notion of probability had to be derived from that "conceived by all of us in everyday life." But first, it was important to identify in the abstract a "logic of the probable," to be grounded on basic assumptions about events.

To do this de Finetti followed the axiomatic approach. He pointed out that when we cannot say precisely which of a set of mutually exclusive events will occur, we can rely on comparisons between events to construct a probabilistic representation of

[14] It is significant that the English translation of de Finetti's essay did not appear until 1964, along with a reprint of Ramsey's article, in a volume devoted to introducing the classics of the subjectivist approach to an international audience. The volume, edited by Henry Kyburg and Howard Smokler (1964), testified both to the growing popularity of the subjectivist approach and to the need to make the hitherto largely unknown foundational writings of the approach available to a wider audience.

the outside world. First, there is a qualitative notion of more, less, or equally probable that allows us to order events according to their probability: if *E'* is considered at least as probable as *E"*, then $p(E') \geq p(E")$. Secondly, there are an impossible event and a certain event, which delimit the range of uncertain events: if *I* is the impossible event and *C* is the certain event, then for each *E,* there must be a probability such that $0 = p(I) < p(E) < p(C) = 1$. Finally, there is a notion of transitivity whereby if one event is more probable than another, and this other one is more probable than yet another, then it must hold that the first is more probable than the third: if *E* is more probable than *E'* and *E'* is more probable than E", then $p(E) > p(E")$.

De Finetti's primary contribution was to demonstrate that, if one interprets the three properties of ordinary probabilistic reasoning as axioms that must hold for the analysis of the objects under scrutiny—the events—then the addition of a further axiom is sufficient to obtain the fundamental result that the 'qualitative' probabilistic evaluations formulated in everyday life actually constitute a 'quantitative' probabilistic measure that respects the property of additivity. De Finetti's fourth axiom—defined as "the essential property on which the entire calculus of probability rests" (de Finetti, 1930, p. 2, author's translation)—stipulates that, given two disjoint events, the order between the two events is preserved if each event is joined by a further event *E* (provided it is also disjoint from the previous ones): if *E'* is more probable than *E"*, then $p(E' \cup E) > p(E" \cup E)$.

This approach has qualitative features that make it very similar to that of Keynes, toward whom, moreover, de Finetti was less critical than Ramsey.[15] But in the light of the fourth axiom, when applied to a set of events that are disjoint from each other and that are judged by the subject to be equiprobable, it becomes possible to advance toward a quantitative understanding of probability. Thus, according to de Finetti ([1937] 1964, p. 101),

> while starting out from a purely qualitative system of axioms, one arrives at a quantitative measure of probability, and then at the theorem of total probabilities, which permits the construction of the whole calculus of probabilities.[16]

However, de Finetti also introduced a more direct way of deriving subjective probabilities in quantitative terms, linked to ordinary reasoning. As with Ramsey, this direct way was to interpret the individual's degree of belief in the occurrence of an event as revealed by the conditions under which the individual would be willing

---

[15] Keynes's *Treatise* was not mentioned in the 1937 article, and only in passing in an earlier philosophical essay in Italian entitled "Probabilism." But in an another Italian article of the following year, later translated as "Cambridge probability theorists"—Keynes and Harold Jeffreys, but not Ramsey, whom de Finetti did not mention until the early 1950s—de Finetti ([1938] 1985) emphasized the fact that the *Treatise* had broken with the English frequentist tradition, thus opening the field to an analysis of the logic of probability based on ordinary reasoning. Although critical of Keynes's treatment of non-numerical probabilities, de Finetti acknowledged that Keynes's qualitative approach had been a fundamental step in enabling the revaluation of an epistemic approach to probability.

[16] De Finetti also introduced a fifth axiom for reasoning about conditional probabilities, that is, probabilities that are derived as a function of the arrival of new information. For the sake of simplicity, both here and in what follows, we will not dwell on the issue of conditional probabilities.

to bet on that event. De Finetti noted that this approach was problematic because it required the employment of monetary sums, which did not appear in the previous abstract reasoning. But since the possibility of deriving probabilities directly from choices made this route easier from an interpretative point of view, de Finetti moved from the axiomatic approach to the pragmatic one of deriving probabilities from actions, with the consequences of the actions being measured in money. In both cases, however, the ultimate goal was to show that it is possible to transform the uncertainty expressed by a degree of belief in the occurrence of an event into a precise number.[17]

Probabilities can be derived simply by asking an individual to indicate the sum he would be willing to exchange for the possession of a given sum S if an event E occurs (and zero if it does not occur). An individual who declares himself willing to make this exchange—the outcome of which is conditional on the occurrence of E—for some maximum sum P, signals that $P/S = p$ is the value at which he would buy the bet, per unit of money. But this also means that p is his degree of belief in the event, or "more simply, that p is the probability of E (according to the individual considered)" (de Finetti, [1937] 1964, p. 102). Thus, abstracting from the utility of monetary amounts, p emerges from the ratio of the (certain) sum that a person initially owns (and which is part of her endowment if she does not bet), to the (uncertain) sum S that the same person would obtain if she bets and wins. It is apparent that this is a ratio that is analogous to $u(y)/u(x)$, that discussed by Ramsey.

It must also be noted that, as in Ramsey, this kind of reasoning applies to an individual who maximizes expected values. In this setting, an expected win is associated with the bet, and the individual who is willing to accept it by paying the price P indicates that he has in mind a probability of winning that derives from the equality $P = pS + (1 - p)0$. Thus, given P and S, the unknown $p = P/S$ is determined.[18]

Once an individual has formulated his probabilities on a set of events in this way, e.g., on the victory of each of the horses participating in a race, two cases are possible: either it is possible to bet with him and certainly gain, or not. According to de Finetti, the individual is "coherent" if the prices he is willing to pay—his valuation of the probabilities—are such that it is not possible to gain certainly if by betting with him: in fact, this consistency condition constitutes "the sole principle from which one can deduce the whole calculus of probability" (de Finetti, [1937] 1964, p. 103). On such a basis, it could be argued that the individual's subjective evaluations are indeed probabilities, in the sense that they can be represented by a mathematical measure that satisfies the properties of a probability.

---

[17] The problem of the utility of money arises again here: in general, as Bernoulli had noted when he introduced the idea of maximizing expected utility, the same amount of money has different utilities depending on the individual's initial wealth when evaluating the bet. De Finetti therefore focused on bets with amounts of money that were small compared to the individual's wealth, an assumption that allowed him to avoid dealing with differences in marginal utilities.

[18] It also follows that, for an individual who values the subjective probability of an event E equal to p, a "fair bet" can be called that which gives one of the competitors a gain $(1 - p)S$ if E occurs and gives the other a gain pS if E does not occur (de Finetti, 1930, p. 3).

It is important to reiterate that the logic followed by de Finetti—exactly as we have seen with Ramsey—entails that probability is subjective, but not arbitrary. On the one hand, it is true that if a person has a degree of belief p in an event E, it is accepted that p can be any value, representing an a priori probability that depends on her personal inclinations, as is normally assumed for preferences over baskets of goods. On the other hand, this person *must* have a degree of belief $1 - p$ in the negation of the event E. And since each probability is inferred from a bet, this means that, if asked at what price she would bet on the negation of E, she should bet at a price revealing $1 - p$. However, neither de Finetti nor Ramsey mentioned the possibility that an individual might decide not to bet on an event. Unlike what Keynes would surely have admitted, de Finetti and Ramsey's ideal person cannot avoid choosing which side of the bet to take at a given price.

Crucially, as we already anticipated in Chap. 3, this condition may seem obvious for objectively given probabilities, but it is stringent when applied to degrees of belief. De Finetti basically assumed that, in order for probability consistency to be guaranteed, once a person has stated the maximum price at which she would participate in the bet offered by a bookmaker, she must also be willing to offer the same bet at the same price to anyone who is willing to participate in it, hence playing the role of the bookmaker. But, as Keynes would argue, it is one thing to give up P for sure in the prospect of being able to win S with a given probability, but quite a different one to face the prospect of losing S with that probability, while receiving P for sure in return.

Thus, the consistency condition is crucial for the argument that sees degrees of belief represented by probabilities to be valid. De Finetti's "logic of the probable" is based on the same argument as Ramsey's "logic of consistency," since both hinge on what we have called the Dutch book argument. It is this further assumption that ensures that the individual is capable of expressing an accurate, definite probability in any decision context. We will see in Chap. 6 that such an assumption would be challenged even by some decision theorists who considered themselves subjectivists, with particular reference to Daniel Ellsberg and his idea that it is possible to identify ambiguous decision contexts in which consistency cannot be considered a rationality condition.

Before concluding by examining what of de Finetti's thinking is relevant to the analysis of uncertainty, it is important to highlight two aspects of his approach. First, de Finetti explicitly adopted an interpretation of pragmatism that would become dominant in the following years. Not only did he assume that degrees of belief, and hence probabilities, could be inferred from actions, but in order to derive probabilities he defined the operations from which they were in practice inferred. According to de Finetti, the notion of subjective probability would be meaningless if it were not possible to specify how probability is actually calculated. In the absence of a clear definition of this procedure—something that, as we saw above, Ramsey believed to be the case with Keynes—one could not be satisfied with the theory of subjective probability.

This approach was established in the exact sciences in the late 1920s in the form of operationalism, usually associated with the American physicist Percy Bridgman

(1927), but was imported into economics only when the use of mathematical methods became widespread. In order to give economics the status of a true science, it was necessary to overcome the imprecise use of concepts that was traditional in political economy, even when economists attempted to develop an argument in mathematical terms. As mentioned in the previous chapter, also Morgenstern considered this as a crucial feature for the development of economic theory.

A second relevant aspect of de Finetti's approach was that, since he suggested to derive probabilities from actual behavior, it seemed obvious to him to argue that the proposed notion of probability was one used "by all men in all the circumstances of life" (de Finetti, [1937] 1964, p. 111). The issue is that of the descriptive capacity of the theory, which de Finetti essentially adopted with this clear-cut consideration. But just as it is well known that people make elementary mistakes in standard logic, one can well imagine that it is even more common to observe real behavior that violates probabilistic logic. An important debate on these aspects opened up in the 1950s, to which we will return in the next chapter. In the end, especially after his encounter with Savage, even de Finetti admitted that his logic of the probable applied to an ideal, not a real, economic subject. When the 1937 essay was translated into English, de Finetti ([1937] 1964, p. 111) clarified the matter by adding in a 1964 editorial note that "probability theory is not an attempt to describe actual behavior; its subject is coherent behavior, and the fact that people are only more or less coherent is inessential." The use that can be made of the theory is therefore normative, not descriptive.

The application of subjective probability theory to economic decisions under uncertainty will find its final formalization in Savage's work, to which the rest of this chapter is devoted. Before proceeding, however, it may be useful to note that, although de Finetti's early work was not as explicitly concerned with utility and its maximization as Savage's later work, de Finetti did clarify some important aspects of the issue. Partly as a result of interests linked to his early work as an actuary, in later years, de Finetti applied his reasoning to the theoretical problem of insurance, proposing an approach that departed from the usual one of identifying a group of analogous cases within which to place the individual insurance contract.

As we have seen, subjectivist reasoning allows a probability to be assigned to any particular event, however singular it may be. Moreover, it is normal for two different parties to assign different probabilities to the same event, even if they have the same information about its possible occurrence. In this context, even the most peculiar insurance contract can theoretically be offered, such as the one concerning the possible capture of the Loch Ness monster mentioned in our discussion of Knight's theory in Chap. 3. The exchange, which in a context of certainty is justified by the different rates of substitution between the two parties involved in it, can now be justified as arising from the different valuations of the probability of the event occurring.

In dealing with the question of insurance, de Finetti noted that the distinction between risk and uncertainty as set forth by Knight had no reason to exist in a subjectivist context. Since every event can be assigned a precise probability, every event is theoretically insurable; with respect to the association between measurability and insurability proposed by Knight, de Finetti's analysis thus implies the opposite of what Knight had argued. Insurance and betting have very different social functions, but the mechanism on which they are based is the same, so that any kind of risky situation, even the most uncertain in the style of Knight, can theoretically find insurance cover.

Having clarified that with de Finetti, as with Ramsey, we are moving toward the negation of any difference between risk and uncertainty that will find its definitive systematization in Savage's work, it should however be noted that there are aspects of de Finetti's analysis of insurance that make him a less dogmatic figure than is usually believed. First, he was aware that there is a problem of information asymmetry in any insurance contract, all the more so given its singularity, and that this requires specific contractual forms, as explored in the microeconomic theory of adverse selection and moral hazard in the 1970s. Second, he acknowledged that in reality there might be situations in which the risk assessments made by different insurers would be fairly consistent, and situations in which this would not be the case. In this second case, de Finetti argued, Knight's distinction might still make sense: for the real insurance market to effectively insure all risks, there must be some uniformity in the subjective assessments of insurers, and if this is lacking, it is not certain that every possible risky situation is actually covered (Feduzi et al., 2014).[19]

More generally, de Finetti was referring to the question of the completeness of markets. In economics, markets are said to be 'complete' if every possible exchange that individuals find advantageous can actually be made, that is, in the case of insurance, if every possible risk can be insured. Contrary to Knight's idea of the different insurability of risk and uncertainty, the subjectivist approach makes it possible to state that this is theoretically always guaranteed: the insurer will set an insurance premium for any type of risk and the contract will be signed if those who wish to insure themselves are willing to pay that premium. But de Finetti's study of the economics of insurance left room for peculiar cases in reality—such as divergence of opinions among insurers—in which market incompleteness manifests itself. Thus, de Finetti admitted, Knight's argument, which had been rejected in the abstract, became relevant again in practice.

It is important to recall that the issue of diversity of opinions is a critical one for the subjectivist position, as we saw above when Keynes pointed it out in his commentary on Ramsey: if probabilities are subjective in the sense of personal, how can we ascertain the 'real,' empirically relevant probability of an event, when two different people assign different probability values to it? As will be discussed in Chap. 6, this question is central in the debate about the extent to which the new subjectivist approach was genuinely general.

[19] In insurance markets, it is common practice for an insurance company to underwrite large risks only on the condition that they can be reinsured with other players. The Lloyd's of London, often cited in the literature on these issues, did not start out as an insurance company, but as a group of individual insurers willing to take risks in partnership. In this case, if there are large differences in the risk assessments of different insurers, reinsurance or risk-sharing may not be warranted, and the risk may remain uninsured.

5.4 Savage and the Bayesian Foundations of Decision Theory Under Uncertainty

We have seen in the previous sections that the notion of subjective probability, in its accomplished form as a measure of degrees of belief that satisfies the calculus of probabilities, originated from the independent intuition of Ramsey and de Finetti. But when vNM developed the theory of expected utility in the 1940s, the probabilities to which they referred to were objectively given, and the application of their theory was justified only when discussing situations of risk, rather than uncertainty. In contrast, both Ramsey and de Finetti placed emphasis on the actions of individuals who make probabilistic judgments about all kinds of events. The two scholars agreed on the premise that knowledge is in essence probabilistic, and that the goal of the sciences that examine human behavior is to study it when the results of actions are not certain. Accordingly, probabilistic judgments, as in Pascal and other thinkers before the advent of frequentism, were associated by Ramsey and de Finetti with the lack, rather than the availability, of knowledge. Nevertheless, they argued that choices should be studied on the assumption that individuals seek to maximize their utility from actions, a premise that was already unanimously accepted in the context of certainty and that vNM had enlarged to cover risk.

The explicit link between subjective probability and the maximization of an expected utility function was introduced in an accomplished manner by the American mathematical statistician Leonard ‘Jimmie’ Savage (1917–1971), whom we have already met in the previous chapters for his role in popularizing the potential of vNM’s expected utility theory in the 1948 essay he co-authored with Friedman. Savage studied mathematics at the University of Michigan, where he obtained a Ph.D. in 1941. After being von Neumann’s assistant at Princeton for a year, he joined a research group at Columbia University that provided statistical analyses to the US Army during World War II. Here he met Friedman, with whom he would later collaborate when he moved to the University of Chicago, where he was one of the founders of the Department of Statistics in 1949. Once in Chicago, Savage found himself involved in discussions on the axiomatization of expected utility, much of which took place at the Cowles Commission for Research in Economics, the aforementioned research center usually regarded as the principal institution that inherited the instances of the Vienna Mathematical Colloquium.

Savage’s work showed the mathematical conditions that made it possible to extend vNM reasoning to situations of uncertainty. This development was later supported by those economists engaged in formal developments in economic theory, as evidenced by a notable debate held in Paris during a conference focused on risk and uncertainty, organized by French critics of the expected utility approach in 1952. In the years following the publication of his major work, the 1954 volume *The Foundations of Statistics*, Savage devoted himself largely to investigating the techniques of so-called Bayesian statistics, in close collaboration with de Finetti (Lindley, 1980).[20]

---

[20] We will return to the meaning of the adjective 'Bayesian' later to identify the subjective approach to probability. It should be made clear at the outset, however, that for decision theorists and

We have noted in the previous chapter that, in his essay with Milton Friedman, Savage was the first author to clarify that the principle of utility maximization is applicable even when considering expected utility, provided that some axioms are applied to choices over lottery. After vNM, the perplexities of the leading marginalist economists of the nineteenth century—motivated by the defense of diminishing marginal utility—could be overcome. The utility function that vNM derived from choices could represent different attitudes to risk, not only aversion to risk—thus an attitude compatible with diminishing marginal utility—but also neutrality toward, or love of, risk—thus a psychological motive that could justify, for example, participation in unfair bets. However, the problem of uncertainty remained to be addressed: vNM dealt only with situations in which probabilities could be derived either from the observation of frequencies or from considerations of symmetry.

In thinking about the matter, Savage came to realize that de Finetti's probabilistic approach provided him with a framework wherein the degrees of belief in the occurrence of any kind of event could indeed be measured. In light of this, it was merely a matter of identifying which axioms would support the conclusion previously reached by vNM about risk on the basis of objective probabilities. In particular, Savage moved in the methodological groove traced by Marschak. This entailed viewing vNM's reasoning as an extension of the axiomatic structure that was already valid for choices under certainty, through the introduction of a further axiom, the so-called independence axiom.

According to Marschak (1950, pp. 111–112), this approach was part of a process aimed at identifying the conditions under which ordinary logic could be applied to any decision context. The objective was to identify the specific "additional definitions and postulated rules" that would make it possible "to 'prolong' logic and arithmetic into the realm of decision." As indicated by vNM, the axioms had to have intuitive meaning so that their reasonableness could be directly assessed. Once a system of axioms had been identified that was both simple and transparent, it would then be possible to provide a structure for the choices of a 'rational' economic individual. The main objective, therefore, was to identify decision rules that could be interpreted as "recommendations to be followed," rather than as descriptions of actual behavior.

Savage endorsed this approach, which identifies decision theory as a predominantly normative scientific discipline. In the introduction to his book, he reiterated Marschak's theme, stating that he wanted to test whether logic could be extended to decisions "by means of principles accepted as those of logic itself" (Savage, [1954] 1972, p. 6). However, the aim now was to extend the approach to cover not only situations of risk but also those of uncertainty. Savage's basic idea was that one could reason in the same way about both 'lotteries'—outcomes that occur with known probability—and 'gambles'—outcomes associated with any kind of event. This approach would allow the characterization of a preference structure that orders

economists, it is not simply a matter of applying Bayes's theorem. Bayes's theorem remains an integral component of the subjectivist idea that there is a precise way in which initial (a priori) probabilities are updated as new information arrives, until final (a posteriori) probabilities are determined. However, the term 'Bayesian' mainly refers to the idea that an initial probability is always precisely assignable to each person in each situation.

consequences that depend on the occurrence of uncertain events of any kind—from the discovery of a new technology to the election of a new US President—about which individuals can only have degrees of belief, since in such cases they cannot rely on the objectively known probabilities of lotteries. In particular, Savage presented an axiomatic framework that allowed the derivation of both of the decision-maker's subjective components—probability and utility—from choices involving uncertain consequences. This constituted a step that neither vNM nor de Finetti had taken, and that Ramsey had delineated without actually providing a theorem proving it.

According to Savage, uncertainty can be treated through the identification of a set of 'states of nature,' called the state space. A state of nature incorporates all the uncertainty associated with a specific decision situation, in the sense that it theoretically describes every aspect relevant to the outcome of the choice; and the list of states is exhaustive, in the sense that all possible relevant issues are covered. But at the moment of making a choice, it is not known which state from a given complete list of states will be realized. To simplify, let us take the example of a horse race in which only three horses participate. There are three states of nature: s_1 indicates the victory of the Horse A, s_2 that of the Horse B and s_3 that of the Horse C. The set $S = \{s_1, s_2, s_3\}$ is the space of states, in which all possible events/results are included. The events are subsets of S: for example, $\{s_1, s_2\}$ is the event in which the race is won by either Horse A or Horse B. In general, one can reason about the set of subsets of *S*, called the algebra of events, but for simplicity one can limit the analysis to the singular events: $\{s_1\}$, $\{s_2\}$, $\{s_3\}$.[21]

In addition to states of nature, Savage identified as 'consequences' (outcomes) a set of possible outcomes x, with $x \in X$, the set of all outcomes. Each consequence is associated, on the one hand, with the decision-maker's choice and, on the other, with the state s_i that occurs. The objects of choice are 'acts,' which at this point are nothing more than a function that associates a consequence with each state of nature. An act f belonging to the set of acts F is such that $f(s) = x$. The choice problem can be represented by a matrix of consequences, with states of nature as columns and acts as rows, as in Fig. 5.2.

The matrix in the figure is easy to read. If the decision-maker bets, for example, on the victory of Horse B he will get: 0 if Horse A wins, 1 if Horse B wins and 0 if Horse C wins. In fact, the structure of the matrix corresponds to that which we deduced from Blaise Pascal's reasoning in Chap. 2. But the setting up of this way of reasoning, that is, the acts-states-consequences scheme, is due to Savage.[22]

When choosing an act f, the individual is evaluating a set of consequences, one for each possible state of nature (each f is in fact a vector). The individual is proposed to

---

[21] As mentioned in Chap. 2, Kolmogorov ([1933] 1950) was the first scholar to give a formal mathematical structure to the notion of probability using an algebra of events. Probability is a measure on events that satisfies the properties of non-negativity, additivity and normalization to 1. Whether this structure is then used to present an objective or a subjective interpretation of probability is a matter of interpretation rather than mathematics.

[22] In fact, Savage transferred to the sphere of economic decisions the reasoning structure that Abraham Wald (1950) had developed for statistical decisions. Wald's influence on Savage's representation is reconstructed in Giocoli (2013).

Fig. 5.2 Savage's matrix of consequences

STATES / ACTS	s_1	s_2	s_3
BET ON HORSE A	1	0	0
BET ON HORSE B	0	1	0
BET ON HORSE C	0	0	1

reason about pairs of acts, so that for each f, $g \in F$ it is possible to identify a preference. In this way, the preference structure is characterized by observing choices, which will obviously depend on the decision-maker's opinion on the final outcome of the horse race, but without assuming as given either the probabilities or the utilities. In the trivial case we are discussing, the choice to wager on, say, Horse B indicates that the subjective probability assigned to state s_2 is greater than that assigned to the other two states.

On the basis of this description of a probabilistic situation, the procedure followed by Savage was to identify the axioms that were needed to prove a theorem analogous to that of vNM. Given a set of basic axioms about how the preferences of a 'rational' individual must be structured in general with respect to a known set of possible acts—then on the pairs f, $g \in F$—Savage managed to prove that there exists a set of non-negative numbers $p(s_i)$—for which it holds that $\Sigma_i\ p(s_i) = 1$, so that they can be identified as (subjective) probabilities—and a utility function $u(x)$ such that each act f is evaluated by means of $\Sigma_i\ p(s_i)\ u(f(s_i)) = \Sigma_i\ p(s_i)\ u(x_i)$. Ultimately, if the preferences expressed by the individual satisfy the set of axioms indicated by Savage, then it can be shown that it is 'as if' he is choosing by maximizing a subjective expected utility function, where the adjective 'subjective' refers to the probabilities of the summation.

The crux of the analysis was the identification of an axiom similar to de Finetti's fourth axiom—the one he had introduced to prove the additivity of the probability function—according to which the order of probabilities is preserved when a third independent event is added to two ordered events. Suppose, as Savage ([1954] 1972, p. 21) argued, that a businessman is considering whether to buy a particular piece of property. His decision depends in general on a number of possible future situations, that is, the consequences in all possible states of nature. But suppose he focuses solely on the effect that the upcoming US presidential elections might have on his purchase, because he is convinced that the outcome might be relevant to the attractiveness of the purchase. So he evaluates whether to buy (act 1) or not to buy (act 2) in case the winner is a Republican candidate (state 1), and decides that he would buy. Then he evaluates whether to buy (act 1) or not to buy (act 2) in case the winner is a Democratic candidate (state 2), and again decides that he would buy. Given this assessment, he decides to buy, even though he does not know what the outcome of the next election will be.

Fig. 5.3 Savage's sure-thing principle

STATES / ACTS	S_1 *E*	S_2 *non-E*
GAMBLE 1	*A*	*C*
GAMBLE 2	*B*	*C*

This principle of behavior is called the "sure-thing principle" by Savage. For Savage, the idea underlying the sure-thing principle was at least as obvious as that underlying the usual axioms on the ordering of preferences under conditions of certainty, which we recall are completeness and transitivity. Indeed, Savage (1954, p. 21) commented: "I know of no other extralogical principle governing decisions that finds such ready acceptance." Such a claim evidently referred to the question raised by Marschak, namely, how to "prolong" logic into the realm of decision. But now it made it possible to admit uncertainty into that realm.[23]

To illustrate this schematically, Savage followed a procedure analogous to that used by vNM to introduce the choice between lotteries. He considered three objects of choice *A*, *B*, *C*, on which a preference structure is given such that, for instance, $A \succ B \succ C$. There are then two mutually exclusive states of nature, *E* and *non-E*, and two acts that represent the 'gambles' that are the objects of choice, which are now not lotteries but more generally gambles since the probabilities are not objectively given. The first gamble has consequence *A* in the first state of nature, that is, if event *E* occurs, and *C* in the second state of nature, that is, if event *E* does not occur. The second gamble has consequence *B* in the first state of nature, that is, if event *E* occurs, and *C* in the second state of nature, that is, if event *E* does not occur, as showed in the matrix in Fig. 5.3.

Since the consequences related to state 2 are the same no matter which act is chosen, they cannot influence the choices. In this simplified version, the sure-thing principle states that the choice of an act should not depend on the situations in which the acts among which the choice is made have the same consequence, but only on the situations in which the acts have different consequences.

As for the fact that the presence of *C* cannot change the structure of the preferences expressed when *A* and *B* are the sole objects of the choice, Savage argued that the choice of the first gamble seemed obvious to him, because by choosing this way "I guarantee that whatever event occurs I will have nothing to reproach myself with." The intuitive appeal of the axiom thus rests on the assumption that only the consequences of a decision matter to its evaluation. Accordingly, a generalized version of the principle states that if the consequences of one act are at least as

[23] A similarly bold statement was used in the very first presentation of the principle: "a principle that we believe practically unique among maxims of wise action in the face of uncertainty, in the strength of its intuitive appeal" (Friedman & Savage, 1952, p. 468).

favorable as those of another in all possible events, and are more favorable in at least one event, then that act is superior overall.

Of course, if this is true in a situation in which the probabilities of the events are only subjectively known, then it must be even truer when the probabilities of the events are objectively known, as in the case of the vNM lotteries. If p is the known probability that E will be realized, then $pu(A) + (1 - p)u(C) > pu(B) + (1 - p)u(C)$ holds. Illustrated in this way, said Savage, the axiom of independence becomes a "special case" of the new principle, which holds even when the probabilities are not objectively known.[24]

The new principle/axiom—borrowed from de Finetti's fourth axiom and used in place of the axiom of independence—allowed for the 'extension' of logic to situations in which the results of acts depend on events about which subjective probabilities are formulated.[25] But whereas de Finetti had not been concerned with utility, Savage worked in a context in which he was able to determine an ordering of acts based jointly on utility and probability. The objects of choice are not necessarily monetary sums or utilities; they could be, for example, A a trip to Paris, B a trip to Madrid and C a trip to Montreal. Regardless of the value in monetary or utility scale that the individual may assign to consequences A, B and C, only the structure of preferences about the consequences is assumed to be known. It is this character of generality that makes Savage's contribution regarded by economists as the foundational one on this matter.

Moreover, we have assumed that $A \succ B \succ C$, but if C were a trip to Rome—such that, for example, the preferences were $C \succ A \succ B$—the reasoning would remain unchanged. Since the trip to Rome is conditional on the non-occurrence of the event, whatever the selected act, it cannot influence the preferences between A and B any more than a non-preferred trip such as the one to Montreal. We shall see in the next chapter that this aspect was the focus of the criticism levied by the few authors who did not accept the alleged possibility of reducing the 'uncertainty' generated by generic events, which we have identified with gambles, to the 'risk' associated with lotteries.[26]

---

[24] This explanation of the axiom is taken from a 1950 correspondence between Savage and Samuelson, reconstructed in Moscati (2016, p. 230). The two were still engaged in a discussion about the meaning of the axiom of independence, which Marschak (1950) had just introduced to clarify vNM's argument. Savage perceived the axiom to be self-evident, whereas Samuelson opposed it because it did not seem to him to contemplate the elements of complementarity and substitutability that are typically allowed in the context of choices under certainty. Samuelson would later admit that it was Savage who persuaded him of the merits of incorporating the independence axiom, designating "the forthcoming book by L. J. Savage … a landmark in the history of probability theory" (Samuelson, 1952, p. 671). But in reasoning on this matter, Savage was actually introducing the axiom that he would later call the "sure-thing principle."

[25] There is a technical reason that induced Savage to designate this generalization as a 'principle' rather than an 'axiom,' but it is too specific to explain here. On this point, see Gilboa (2009), which introduces the reader to the mathematical details of the whole Bayesian approach.

[26] It is interesting to mention how Savage, in a 1961 private correspondence with de Finetti, reconstructed the way he had arrived at his result. Savage recalled that at a seminar in 1949 he had presented his ideas, still in preliminary form, to colleagues at the Cowles Commission. Tjalling

In conclusion, in the vein of vNM, Savage presented a representation theorem: he proved that individuals whose preferences satisfy a set of axioms would always choose as if they were maximizing some subjective expected utility. Savage proposed a complex mathematical proof, where more technical axioms were added to the axioms already mentioned, and indeed in modern textbooks, his proof has been replaced by the version of Anscombe and Aumann (1963), who simplified the argument by assuming that the consequences of an act are lotteries with known probabilities. Nevertheless, Savage's result is unanimously regarded as the crucial step in definitively establishing the legitimacy of expected utility maximization in economic theory.[27]

5.5 Concluding Remarks

Savage's work represents the culmination of the subjectivist revolution in the field of decision theory. We have seen in the previous chapters that, following the so-called Paretian revival, economics began to treat the decision-maker as rational on the basis of hypothetically observed choices, and thus to focus on preferences rather than utility. This was despite the fact that optimal choices were still represented in terms of the maximization of a utility function, a representation that was retained for reasons of analytical simplicity and convenience.

Indeed, as noted in the previous chapter, both John Hicks's representation—which used the marginal rate of substitution to quantify utility gains—and Paul Samuelson's—which showed how actual choices could be judged to be more or less consistent—no longer started from a utility function. Rather, a utility function was derived from preferences over bundles of goods, whether assumed (Hicks) or revealed (Samuelson). The subsequent extension by vNM to risky choices, while crucial to the rehabilitation of Bernoullian expected utility, constituted an intermediate step, because it was still tied to the objective component represented by the known probabilities of the random mechanism conditioning outcomes. The final result achieved by Savage, following in the footsteps of de Finetti and Ramsey,

---

Koopmans, then director of the research center, was so enthusiastic about the possibility of bringing uncertainty within the canons of expected utility that he asked Savage to schedule a more thorough presentation as soon as possible. To prepare the new presentation, Savage locked himself in the library to search for possible sources for his work. Savage revealed to de Finetti that he had then found himself studying the paper de Finetti had published in the *Annales de l'Institut Henri Poincaré*, and that in fact the new presentation turned out to be a kind of review of that paper. This letter is quoted in Zappia (2024).

[27] To Anscombe and Aumann (1963) is also traced the suggestive distinction that while vNM had dealt with 'roulette lotteries'—with the probabilities of success linked to a random mechanism such as a spin of a (unrigged) roulette wheel—thanks to Savage one could also study 'horse lotteries'—with the probabilities of success linked to the valuations of individual bettors.

brought probabilities back into the realm of individual subjectivity, with rationality understood simply as the consistency of degrees of belief.

Savage's step was a crucial one in the characterization of economics that emerged in later years. The extension of logic to decision-making, as advocated by Marschak and the mathematical economists at the Cowles Commission, was embodied in the identification of which axioms would allow the attribution of probability properties to individual degrees of belief. This step completed the process of transforming economics from a science that studied maximizing individuals to one that studied coherent individuals: the idea of rationality as consistency of choices finally found application to decision contexts where probabilities were 'personal,' as Savage specified. According to this approach, economic theory can provide a precise solution even in the most complex cases. This is the case even when choices are made in situations where knowledge is considerably lacking, such as those identified by Keynes as "very uncertain" in the final part of the quote with which we opened this volume. With Savage, the maximization of expected utility—now calculated in terms of both the subjective components of choice, utilities *and* probabilities—became an axiomatically grounded criterion and therefore of general significance, with the same status as the general rules offered by the exact sciences. In the words of one of the most prominent theorists of our time, Savage's representation theorem can be regarded as "the crowning glory of choice theory" (Kreps, 1988, p. 120).

The theory of rational choice that consolidated in the following years—initially referred to as 'Bernoullian' due to its revaluation of Bernoulli's seminal contribution—is indeed an all-encompassing theory of economic decisions. It can be applied to any decision-making context, so much so that it would soon be referred to as 'Bayesian' decision theory. Bayesian because it assumes that economic agents are able to express their opinions about each possible future event that may condition the outcomes of their actions as definite, precise numerical probabilities. This conceptual framework enables the comparison between economic agents' degrees of belief and the objective data that emerge, and then the derivation of a posterior probability through the application of Bayes's theorem. But, following the contributions of Ramsey, de Finetti and Savage, decision-makers in economics are Bayesian, in the first place, because of their supposed ability to always assign a prior probability to events, regardless of how complex the decision situation may be. This aspect was absent from Bernoulli's analysis, but it was inherent to the definition of probability as set forth by Bayes and explicitly articulated by Ramsey. It is on this assumed ability to always assign a precise number to the probability of events occurring that the entire Bayesian construction hinges (Gilboa et al., 2008).

This aspect is also emphasized by the game-theoretical reading that became common in later years. Following Savage's scheme, the states of nature can be understood not only as a representation of uncertainty dictated by Nature, but also as the choices of an opposing player. Subjective probabilities about states can well represent the beliefs formulated about the behavior of the other player, that is, the probability with which one player assumes that the other will choose a given strategy. Since outside of two-player zero-sum games, there is no reason to assume that one player should believe that the other would choose the strategy that would hurt him the most—as vNM had assumed with their proposal to concentrate on the max-min—the act (strategy) chosen is the one that maximizes expected utility relative to the probabilities assumed about the other player's act (strategy). We saw in Chap. 4 that the fact that vNM did not consider this issue was criticized by Ellsberg; but now that the player's subjective probabilities are identified, it is possible to make it explicit that uncertainty in the game is related to the behavior of the other player. After Savage, it became possible not only to make assumptions about the behavior of the other player, but also to model the uncertainty inherent in the strategic situation as a set of probability functions, each corresponding to a hypothetical 'type' of opponent player, as suggested by John Harsanyi (1967) in one of the most advanced developments of game theory in the 1960s.

And even the theory of rational expectations itself, usually remembered for its impact on macroeconomic modeling and its implications for economic policy, is but a peculiar application of subjective expected utility theory. Conventional post-World War II macroeconomic models had begun to assume the existence of a probability distribution of the aggregate outcomes, determined by the interactions in the system among economic agents maximizing their own subjective expected utility function. A model with rational expectations is a stochastic macroeconomic model in which it is assumed that the agents—not just the theoretician or the policymaker—somehow get to know the 'true' probability distribution with which aggregate outcomes are realized. It is thus a representation of the economy in which the extreme assumption is made that all decision-makers are represented as if their degrees of belief were summarized by a probability distribution that correctly anticipates the probability distribution of aggregate outcomes generated by their actions (Blume & Easley, 2008).

It can therefore be argued that both game-theoretical developments and the macroeconomics of rational expectations exploited to a high extent the idea of rationality that Savage made possible to employ. However, a theory is only valid according to the axioms on which it is based, so it must be possible to conduct an analysis of the meaning of these axioms, as Friedman and Savage (1948) had already pointed out. The question of whether the axioms of a theory should be realistic or not, to be evaluated in terms of their descriptive appeal or the accuracy of the ensuing predictions, was open to debate. However, as we shall see in the next chapter, the debate focused on the former aspect, and in particular on the meaning of the axiom of the sure-thing. Moreover, since a standard probability measure must add up to 1, the events to which the subjective probability measure is assigned must be part of an exhaustive list, that is, a list that theoretically contains all possible alternatives that the decision-maker has to face. Namely, the uncertainty under consideration cannot contain 'surprises,' that is, unexpected events to which no probability has been assigned.

Obviously aware of the restrictive conditions he was dealing with, Savage ([1954] 1972, p. 83) pointed out that the theory he was proposing seemed to him accurate for a "small world," but that it would be "preposterous" to apply it in a "grand world." Setting up a decision problem with the precision envisioned by the acts-states-consequences scheme is indeed feasible in many cases, and many others can be reconsidered in such a manner that they can be understood in the terms of that scheme, hence as small worlds. However, Savage noted, asserting that this scheme can be applied to any kind of decision would be to make an "extreme idealization." If it were possible to be rational in the grand world, it would be possible to think of a person's entire life as if reduced to a single, all-comprehensive decision.[28]

Although explicitly formulated as the theory of a rational—as ideal—decision-maker, and therefore conceived with normative rather than descriptive intent, the theory that was to become dominant was clearly extreme even on normative grounds. Nevertheless, its potential application to every area of decision-making rendered subjective expected utility theory well-suited for further enhancing the prestige of economic theory. Both the scope of application and the reservations that this approach raised will be discussed in the next chapter.

References

Anscombe, F. J., & Aumann, R. J. (1963). A definition of subjective probability. *Annals of Mathematical Statistics, 34*, 199–205.

Bayes, T. ([1763] 1958). An essay towards solving a problem in the doctrine of chances. *Biometrika, 45*, 296–315.

Blume, L. E., & Easley, D. (2008). Rationality. In L. E. Blume, & S. N. Durlauf (Eds.), *The new Palgrave: A dictionary of economics* (2nd Ed.). Palgrave Macmillan.

Borel, É. ([1924] 1964). Apropos of a *Treatise on probability*. In H. E. Jr. Kyburg, & H. E. Smokler (Eds.), *Studies in subjective probability*. Wiley.

Bridgman, P. W. (1927). *The logic of modern physics*. Macmillan.

Davis, J. B. (2024). *Identity, capabilities, and changing economics*. Cambridge University Press.

de Finetti, B. (1930). Fondamenti logici del ragionamento probabilistico. *Bollettino Dell'unione Matematica Italiana, 5*, 1–3.

de Finetti, B. ([1937] 1964). Foresight: Its logical laws, its subjective sources. In H. E. Jr. Kyburg, & H. E. Smokler (Eds.), *Studies in subjective probability*. Wiley.

de Finetti, B. ([1938] 1985). Cambridge probability theorists. *The Manchester School, 53*, 348–363.

de Finetti, B. (1975). *Theory of probability* (Vol 1). Wiley.

Düppe, T., & Weintraub, R. E. (2014). Siting the new economic science: The Cowles Commission's activity analysis conference of June 1949. *Science in Context, 27*, 453–483.

Feduzi, A., Runde, J., & Zappia, C. (2014). De Finetti on uncertainty. *Cambridge Journal of Economics, 38*, 1–21.

[28] To specify that Bayesian decision theory can be truly valid only in a small world, Savage ([1954] 1972, p. 16) illustrated his distinction between 'small' and 'grand' worlds by quoting the proverbs 'Look before you leap' and 'Cross that bridge when you come to it.' Decision-makers are in a small world when they can always look before they leap. They are in a grand world when there are some bridges that they cannot decide whether to cross before they come to them. Whether this distinction is really apt to address the open-ended world perspective of some critics of the Bayesian approach is questioned in Davis (2024, pp. 48–51).

Friedman, M., & Savage, L. J. (1948). The utility analysis of choices involving risk. *Journal of Political Economy, 56*, 279–304.
Friedman, M., & Savage, L. J. (1952). The expected-utility hypothesis and the measurability of utility. *Journal of Political Economy, 60*, 463–474.
Galavotti, M. C. (1989). Anti-realism in the philosophy of probability: Bruno de Finetti's subjectivism. *Erkenntnis, 31*, 239–261.
Galavotti, M. C. (2001). Subjectivism, objectivism and objectivity in Bruno de Finetti's Bayesianism. In D. Corfield & J. Williamson (Eds.), *Foundations of Bayesianism*. Kluwer.
Galavotti, M. C. (2011). The modern epistemic interpretations of probability: Logicism and subjectivism. In D. M. Gabbay, S. Hartmann, & J. Woods (Eds.), *Handbook of the history of logic* (Vol. 10). North-Holland.
Gaspard, M. (2014). Logic, rationality and knowledge in Ramsey's thought: Reassessing "human logic." *Journal of Economic Methodology, 21*, 139–157.
Gerrard, B. (1992). From *A treatise on probability* to the *General theory*: Continuity or change in Keynes's thought? In B. Gerrard, & J. Hillard (Eds.), *The philosophy and economics of J. M. Keynes*. Edward Elgar.
Gilboa, I. (2009). *Theory of decision under uncertainty*. Cambridge University Press.
Gilboa, I., Postlewaite, A. W., & Schmeidler, D. (2008). Probability and uncertainty in economic modeling. *Journal of Economic Perspectives, 22*, 173–188.
Giocoli, N. (2013). From Wald to Savage: *Homo economicus* becomes a Bayesian statistician. *Journal of the History of the Behavioral Sciences, 49*, 63–95.
Hájek, A. (2009). Dutch book arguments. In P. Anand, P. Pattanaik, & C. Puppe (Eds.), *The handbook of rational and social choice*. Oxford University Press.
Harsanyi, J. C. (1967). Games with incomplete information played by "Bayesian" players, Part I. The basic model. *Management Science, 14*, 159–182.
Keynes, J. M. ([1933] 1972). Frank Ramsey. In Id. *Essays in biography. The collected writings of John Maynard Keynes* (Vol. X). Macmillan.
Keynes, J. M. ([1937–38] 1979). Letters to Hugh Townshend. In Id. *The General theory and after: A supplement. The collected writings of John Maynard Keynes* (Vol. XXIX). Macmillan.
Kolmogorov, A. N. ([1933] 1950). *Foundations of the theory of probability*. Chelsea Publishing.
Kreps, D. (1988). *Notes on the theory of choice*. Westview Press.
Kyburg Jr., H. E. (2003). Are there degrees of belief? *Journal of Applied Logic, 1*, 139–149.
Kyburg Jr., H. E., & Smokler, H. E. (1964). Introduction. In Id. (Eds.), *Studies in subjective probability*. Wiley.
Lindley, D. V. (1980). Savage. His work in probability and statistics. *Annals of Statistics, 8*, 1–24.
Luce, D., & Raiffa, H. (1957). *Games and decisions*. Wiley.
Marschak, J. (1950). Rational behavior, uncertain prospects, and measurable utility. *Econometrica, 18*, 111–142.
Misak, C. (2020). *Frank Ramsey. A sheer excess of powers*. Oxford University Press.
Moscati, I. (2016). Retrospectives: How economists came to accept expected utility theory. The case of Samuelson and Savage. *Journal of Economic Perspectives, 30*, 219–236.
Nagel, E. (1939). *Principles of the theory of probability*. University of Chicago Press.
Newman, P. (1987). Ramsey, Frank Plumpton (1903–1930). In J. Eatwell, M. Milgate, & P. Newman (Eds.), *The new Palgrave: A dictionary of economics* (Vol. 4). Macmillan.
Ramsey, F. P. (1927). A contribution to the theory of taxation. *Economic Journal, 37*, 47–61.
Ramsey, F. P. (1928). A mathematical theory of saving. *Economic Journal, 38*, 543–559.
Ramsey, F. P. ([1931] 1964). Truth and probability. In H. E. Jr. Kyburg, & H. E. Smokler (Eds.), *Studies in subjective probability*. Wiley.
Runde, J. (1994). Keynes after Ramsey: In defense of *A treatise on probability*. *Studies in the History and Philosophy of Science, 25*, 97–121.
Samuelson, P. A. (1952). Probability, utility, and the independence axiom. *Econometrica, 20*, 670–678.

Savage, L. J. ([1954] 1972). *The foundations of statistics.* Dover Publishing.
Wald, A. (1950). *Statistical decision functions.* Wiley.
Weintraub, R. E. (2002). *How economics became a mathematical science.* Duke University Press.
Zappia, C. (2024). Uncertainty goes mainstream: Savage, Koopmans and the measurability of uncertainty at the Cowles Commission. *European Journal of the History of Economic Thought, 31*, 375–399.

Chapter 6
Uncertainty as Ambiguity: Ellsberg and the Paradoxes of Decision Theory

6.1 Introduction

The story outlined in the previous two chapters may appear to be that of a linear development of economic theory. It may sound like a gradual movement toward the 'right' answer to a problem—the application of choice theory to conditions of uncertainty—that economists had always had well in mind but had long treated unsystematically. Following the approach suggested by mathematically trained decision theorists, who were convinced that general propositions can only be stated as theorems derived from axioms, a general solution emerged in the late 1940s and early 1950s. Decision theory, first under risk and then under uncertainty, consolidated around the idea of consistency, following a trend that had begun to emerge in the discussion of the representation of preferences by indifference curves and then put at the center of their analysis by Frank Ramsey and Bruno de Finetti. This shift was indeed understood by some of the young economists of the time as a necessary evolution, as the decisive step toward an increasing ability of economics to solve problems with the formal mathematical precision characteristic of the exact sciences. Among the advocates of this axiomatic approach were future Nobel Prize winners such as Kenneth Arrow, Gérard Debreu, Tjalling Koopmans and Paul Samuelson.

This is indeed the story often told in traditional textbooks: on the basis of new mathematical tools and their increasingly accurate application, the latest economic theories solve problems left unsolved by earlier theories. The solution to complex economic problems would be found through a use of mathematics that no longer aims merely to provide a synthetic representation of an already intuitively formulated reasoning, as theorized by Alfred Marshall, but to derive results that could only be achieved through the application of increasingly sophisticated tools. In a 1957 volume on the new methodologies that sealed the decade opened by the work of John von Neumann and Oskar Morgenstern, Koopmans offered precisely this reconstruction of the progress that he believed had finally been made in economics. The

C. Zappia, *Uncertainty in Economics*, Springer Studies in the History of Economic Thought, https://doi.org/10.1007/978-3-031-91221-4_6

new mathematical tools enabled the overcoming of an approach that was considered too empirical, that is, based on measurement without theory.[1]

The main demonstration of the fruitfulness of this process of further mathematization through axiomatization came from general equilibrium theory. In a chapter of his book devoted to "Resource allocation and the price system," Koopmans (1957, p. 5) noted that it was the recent introduction of "mathematical tools of a more basic character" into economics that had made it possible "to perceive with greater clarity and express in simpler terms the logical structure of important parts of economic theory." Crucially, with regard to the equilibrium allocation of an entire economic system, it was finally possible to identify with analytical precision the conditions under which there exists a vector of prices of goods that guarantees that the choices of a number m of subjects (consumers and firms)—determining the aggregate demand and aggregate supply of each of a number n of goods—are not only optimal from an individual point of view, but also compatible at the aggregate level, that is, equilibrium choices. In this context, it was shown that the equilibrium between market supply and demand for each good, determined at these prices, has the property of identifying an efficient allocation of goods among subjects, in that it can be improved for some individuals only at the expense of some other individuals.

The First Theorem of Welfare Economics provided compelling demonstration of this efficiency property of general equilibrium. Propositions about the market mechanism, initially articulated by Adam Smith and subsequently developed mathematically by Léon Walras and Vilfredo Pareto, remained unproven until the seminal contributions of Arrow and Debreu (1954). Using the tools employed by Arrow and Debreu—elementary concepts and theorems borrowed from the mathematical theory of linear spaces rather than the traditional tools of differential calculus—the new developments made it possible to "overcome some of the defects of earlier formulations" (Koopmans, 1957, p. 43).[2]

However, the most obvious example of the quality of the axiomatic approach was Leonard Savage's contribution. In another chapter of his volume, entitled "The construction of economic knowledge," Koopmans (1957, pp. 142–143) argued that economic theory should be viewed as a "sequence of models," in the sense that each model—defined by a set of postulates/axioms which represent the essential aspects of the reality that the economist wishes to study—can be regarded as a prototype for

[1] "Measurement without theory" is the title of an essay in which Koopmans (1947) presented the new econometric tools in use at the Cowles Commission in controversy with the empirical analyses of the business cycle in force at the National Bureau of Economic Research, another research center, at the time still under the influence of institutionalist scholars such as Arthur Frank Burns and Wesley Mitchell. Only a precise, theoretical definition of the variables to be studied, and of the relationships between them, would guarantee a proper empirical investigation, in application of the methodology of logical positivism (Boumans & Davis, 2010, pp. 38–41).

[2] Moreover, the very possibility of using the utility function as an index to represent preferences over baskets of goods, to which we have alluded several times in previous chapters, found its definitive axiomatic justification only with Debreu (1959). Debreu specified in what sense the continuity of the set over which choices are made is a necessary condition for a complete ordering over the objects of choice to be synthetically represented by a continuous utility function.

subsequent models that aspire to be more realistic, though probably more complicated. A given model cannot be criticized for being unrealistic, since any model must be based on axioms that are in some way necessarily unrealistic. Nevertheless, the task of economists should be to develop models that are based on the "perception of additional aspects of reality" that are not included in the original model. Koopmans's idea was that the confrontation with reality had to pass through the search for new axioms, because only the axiomatic method would make it possible to determine "which conclusions or recommendations depend on which postulates, and which postulates depend for their validity on which verifications of their implications by accumulated evidence" (Koopmans, 1957, p. 142).

According to Koopmans, the decision theory axiomatized by von Neumann and Morgenstern for situations of risk had become a decision theory for situations of uncertainty because Savage had been able to identify the axiomatic structure that allowed the transition to an equally precise but more realistic model. In particular, the so-called sure-thing principle could be considered a generalization of the axiom of independence. Savage's sure-thing principle made the theory applicable to any kind of gamble, suggestive of situations where the probabilities are not objectively known, hence certainly more realistic than von Neumann and Morgenstern's lotteries.

Koopmans considered Savage's theory to be the best demonstration of the fact that the axiomatic approach was progressive: an aspect of reality that poses a problem for an established model—uncertainty, for the expected utility model designed for risk—is incorporated into a new model through the employment of a new axiom that allows that aspect to be analyzed precisely. Of course, whether an axiom is considered more or less plausible depends primarily on the choice of the theorist. But having shown that maximizing expected utility is equivalent to satisfying certain conditions of coherence—such as that imposed by the axiom of the sure-thing—made it possible to evaluate the meaning of the idea of maximization by checking whether these conditions were satisfied, in a similar way to what Samuelson had proposed for choices under conditions of certainty when he introduced the axiom of revealed preferences.[3]

Pace Koopmans, the lengthy reconstruction of Frank Knight's and John Maynard Keynes's thought that we proposed in Chap. 3 makes it clear, however, that the history of uncertainty in economics that we intend to delineate here is not that of a progressive advance toward a supposedly 'better' theory. This chapter is therefore devoted to illustrating the standpoint of those authors who, during the years of consolidation of what would soon become the dominant theory, continued to resist the idea that uncertainty could be reduced to risk. While accepting in principle

[3] The methodological approach proposed by Koopmans was presented to counter the much better known approach of Friedman (1953), who had argued that theoretical economists should not bother to employ unrealistic axioms if they lead to correct predictions. According to Blaug (1997, Ch. 15), although Friedman's idea was in principle better suited to the increasingly sophisticated developments of the new mathematical models, the approach actually followed by much of later economic theory was that of Koopmans. Indeed, experimental economics—the study of subjects' choices in controlled laboratory settings—was born in those years also with the aim of assessing whether assumptions about risk and uncertainty were more or less realistic.

the subjectivist approach developed on an axiomatic basis summarized by Savage's work, these authors criticized its unconditional application to contexts of choice whose complexity is reminiscent of that highlighted by Knight and Keynes. After an initial section summarizing the theoretical and institutional aspects that favored the reception of Savage's model, we turn to the critical perspective introduced by Maurice Allais and, in particular, Daniel Ellsberg. The paradoxes of choice raised by Allais and Ellsberg are the subject of Sect. 3, while Sect. 4 focuses on the importance of Ellsberg's contribution for the rehabilitation of the distinction between risk and uncertainty, with particular reference to the notion of uncertainty as ambiguity. As usual, the final section provides some concluding remarks.

6.2 Bayesian Philosophy and the Application of Subjective Expected Utility Theory

We have already emphasized that the development proposed by Savage should not be considered simply as the contribution of a single author who brought the theory of subjective probability to its final affirmation. Rather, it was the much sought-after result of the efforts of a group of scholars who shared a commitment to the axiomatic approach, and who, on that basis, sought to broaden the scope of decision theory. It was therefore important to clarify, as Arrow noted in an important review article in 1951, that several alternative approaches to decision theory were considered possible at the time.

Not only were axioms being studied for risky conditions with known probabilities, in the style of von Neumann and Morgenstern, but there were also axiomatically based studies for conditions of 'complete ignorance,' that is, where no probability—not even a subjective one—was assumed to be available to the decision-maker. In the absence of any plausible assignment of probability to the events that might condition the choices, one could imagine using the so-called max-min criterion. Devised by von Neumann and Morgenstern for the study of zero-sum games, at Abraham Wald's (1945) suggestion the max-min could be used to study an individual who makes decisions with a pessimistic approach, and who therefore focuses solely on the worst possible consequence associated with each action, as if facing a malevolent Mother Nature.

But the opposite approach was also studied, representing an individual who, in the same situation, has an optimistic attitude toward the uncertain environment and therefore focuses only on the best possible consequences: this was called the max-max criterion. Together with Leonid Hurwicz, Arrow himself studied what axiomatic conditions could justify a decision criterion that combined pessimism and optimism by means of a formula of the type $\alpha \max_A + (1 - \alpha) \min_A$, according to which for each possible action A only two possible consequences had to be taken into account, the best ($\max_A$) and the worst ($\min_A$). The way in which these two consequences were weighted by the parameter α would represent whether the individual was more or less optimistic about which outcome would be realized. In the absence of probabilities, one had to limit oneself to this type of reasoning, which, although based on axioms, did not seem to have the character of generality of a probabilistic setup.[4]

Finally, Arrow recalled in his review, there were also scholars who rejected a probabilistic approach to decision theory as a matter of method, such as the English economist George Shackle (1903–1992). A peculiar figure who devoted most of his research to delve into the uncertainty aspects of Keynes's *General Theory*, Shackle (1949) proposed to reason not in terms of degrees of belief, but in terms of degrees of disbelief, which he called "potential surprise." To deal with true uncertainty, Shackle rejected the idea that it would always be possible to exhaustively identify the list of possible events, and concentrated instead on the possibility that completely unexpected, not just improbable, events might occur. From this, he noted, follows the technical impossibility of representing degrees of belief through an additive function such as probability: individuals' decision weights cannot cover the entire state space, since it is by definition not fully representable. According to Shackle, only such a description could be used to accurately analyze the expectations of entrepreneurs making investment decisions. The uncertainty he had in mind was the extreme uncertainty of Keynes in the 1937 article we analyzed in Chap. 3, somewhat akin, in Shackle's view, to complete ignorance.

In order to assess the outcomes associated with choices, Shackle therefore proposed the employment of a potential surprise function, somewhat reciprocal to probability. This would allow for the consideration of a "residual hypothesis"—representing events not anticipated by the decision-maker—without reducing the weight assigned to those already considered. The decision criterion proposed by Shackle required that, for each possible choice, only one positive value—the outcome that most attracted the decision-maker's attention among the positives associated with a choice—and one negative value—the outcome that most attracted attention among the negatives—be considered. The two values were then evaluated together with the potential surprise associated with them. Shackle's idea was that the decision-maker needed to consider only these two "focus values" because in a context where not all potentially occurring events are known—that of surprise—it would make no sense to calculate an average of the utilities of the consequences.

In his attempt to justify the employment of an alternative function to probability, Shackle also reasoned about what axioms might justify it. But the axioms of his theory were not considered plausible by Arrow. Despite the attention it received in the early 1950s among scholars interested in the viability of a formally structured alternative to theories of behavior describing uncertainty through probability, Shackle's idiosyncratic attempt was soon marginalized among decision theorists (Zappia, 2014). That

[4] If the parameter α is a subjective variable that characterizes the individual's attitude toward the uncertain environment, the Arrow-Hurwicz formula associates a totally pessimistic attitude, as in Wald, if $\alpha = 0$, and a totally optimistic one if $\alpha = 1$. Any other value indicates a mixed attitude, so that every possible attitude in a situation of complete ignorance can be described. As will be discussed in the next chapter, a generalization of this way of ranking alternatives would emerge as a possible decision rule in the 1980s, when the study of non-additive probabilities became an active field of research among decision theorists.

his theory could be considered as an alternative to the then-consolidating mainstream ended up as an argument put forward only among post Keynesian scholars, to whom we will return later.

However, although this diverse series of studies illustrated novelties in the axiomatization of choice theory worthy of consideration, Arrow concluded his state-of-the-art examination by claiming that most of the issues addressed by these alternative approaches already had a solution. Even if only on the basis of a theory of degrees of belief still in the making, it was possible to conclude that, even under uncertainty, the maximization of expected utility could be regarded as the fundamental criterion. For the subjectivists, the need to study complete ignorance in non-probabilistic terms was not justified: in their view, it could be assumed that the decision-maker would always be able to give a numerically precise prior probability in any situation, however uncertain. Even before Savage's work had been published, and only on the basis of the first notes circulated among his colleagues at the Cowles Commission, Arrow claimed that the research undertaken by Savage could clarify in axiomatic terms the meaning of using subjective probabilities for decision-making. He therefore concluded: "Knight's uncertainty seems to have surprisingly many of the properties of ordinary probability, and it is not clear how much can be gained from the distinction [between risk and uncertainty]" (Arrow, 1951, p. 417). In a sense, the revolution of considering all uncertainties as risks was taken for granted even before it was made public.[5]

In Arrow's reconstruction, as well as in Koopmans's assessment, one can see that there is an appreciation of the fact that decision theorists were finally being able to propose a clear, general decision criterion for conditions of uncertainty. This is the spirit in which economists read Savage's work in those years. Despite the fact that the Bayesian approach remained opposed by a large majority of statisticians in the following years—with Savage committed to defending his theory on this front against supporters of the frequentist position—economists largely embraced Savage's proposal as capable of simplifying the analysis of complex problems that had traditionally been considered impossible to tackle systematically.

Of great importance in this respect was the application of the new methodology to management problems, which followed the expansion of business schools in the United States after World War II. In a 1961 volume entitled *Statistical Decision Theory*, joining efforts with another American mathematical statisticians Robert

---

[5] It is worth noting that in this work, still cited as the best review of the theories of the time, Arrow also discussed Keynes's probability theory, something that Savage will also do in his 1954 volume. Although both Arrow and Savage followed Ramsey and de Finetti in arguing for a strictly subjectivist approach, thus refusing to give behavioral significance to Knight's distinction, they emphasized Keynes's refusal to measure subjective probabilities precisely. Arrow (1951, p. 416), in particular, pointed out the historical importance of Keynes's opposition to the frequentist position, and placed Keynes's theory among those who argued that probability should always be used, even when substantial information is lacking. Keynes's peculiarity was that he belonged to a group of authors who argued that "probabilistic statements alone are not sufficient," not that they were inapplicable. But Arrow's conclusion, later echoed by Savage ([1954] 1972, p. 39), was that the axioms that could justify Keynes's arguably more realistic position were insufficient to guarantee the consistency of choice on which probabilistic approaches had to be based.

Schlaifer, Howard Raiffa argued that the time had come for a systematic approach to business problems and that this could only be achieved by applying the acts-states-consequences scheme proposed by Savage, possibly in the simplified version of a decision tree. In a decision tree, the manager's options could be disentangled into 'nodes' where the manager makes a decision and nodes where the exogenous uncertainty determines which of a set of consequences will be realized, with the knowledge available to the manager at each node—the so-called information sets—well highlighted. The main contribution of Raiffa and Schlaifer's volume was to show how a strictly theoretical work such as Savage's could be applied to the field of managerial decision-making. After Savage, rather than relying on unjustified ad hoc behavioral rules, it was possible to represent estimates as subjective probabilities, therefore allowing for the fundamental step of simplifying the decision-making context, that is, to make decisions on the basis of the maximization principle in any environment.

Business academics should not be worried by the excessive degree of abstraction of these developments since, as Raiffa and Schlaifer clarified in a joint paper with John Pratt, the underlined philosophy was simple:

> First, we do not assume that the decision-maker has established his preferences between all conceivable pairs of lotteries by direct, intuitive evaluation; and second, we do not assume even as regards those lotteries which he has evaluated intuitively that his intuitive evaluations will agree with our … principles of consistent behavior. On the contrary, we ask the decision-maker whether he would like his decisions in complex problems to be consistent in the sense of these … principles with his preferences and judgments as expressed after careful consideration of very simple problems. All that we have to offer the decision-maker is a method of calculation that will insure this kind of consistency; ideal decision-makers whose intuitive behavior is perfectly consistent have no use for our services. (Pratt et al., 1964, pp. 356–357)

Raiffa and Schlaifer's volume was the first in a new series published by the Harvard Graduate Business School called *Studies in Managerial Economics*. The series, according to the editor's foreword, was designed to identify those developments that could enable "business administrators to make wiser decisions in some phase of their activity," and this could be done on the basis of "the application of mathematics and statistics to business problems of decision under conditions of uncertainty" (Raiffa & Schlaifer, 1961, p. v). At Harvard, an attempt was being made to give management education a more quantitative flavor than had traditionally been the case. Teaching based on courses taught mainly by professionals with no specific academic training was gradually replaced by teaching focusing on quantitative methods in courses taught by economists, mathematicians and statisticians, all trained in Ph.D. programs.

For a significant proportion of these academics, therefore, the Bayesian methodology was a way to break away from traditional methods such as hypothesis testing, which were not applicable to some of the more complex business decisions, those with unique characteristics and no past frequency observations. This emphasis on quantitative methods was about to spread to the major North American business schools. By 1957, for example, the Graduate Business School in Chicago was under the direction of Allen Wallis, who had co-founded the new Statistics Department with Savage in 1949. Under Wallis, what is remembered as the 'Chicago approach' to management education was established, based precisely on the application of new statistical methods. Similarly, the subsequent outgrowth of Bayesian game theory and its application to problems of asymmetric information, mentioned at the end of the previous chapter, was largely due to the research of scholars active in the business schools of universities such as Stanford and Northwestern, among whom future Nobel laureates Robert Wilson, Paul Milgrom and Roger Myerson.[6]

6.3 The Paradoxes of Allais and Ellsberg

The kind of decision theory that was to dominate for decades consolidated in the context just reconstructed, around 1960. Its use in economic models is still widespread, and it remains part of the standard core of microeconomic textbooks. Yet, since the early years of its diffusion, there have been signs that the systematization proposed by Savage, while certainly sound, was not really unanimously accepted by economists. Indeed, the most significant of these signs was already present in Savage's own work, as he devoted a section of his volume to addressing potential critiques of expected utility theory, particularly those raised by the French economist Maurice Allais.

We have already mentioned in the preceding chapter that Savage and his American colleagues had defended expected utility theory at the International Conference on Risk and Uncertainty held in Paris in 1952. At that conference, Savage had presented for the first time in detail the axiomatic core necessary to develop a theory of subjective expected utility, introducing the axiom that could play the role of the axiom of independence in the case of unknown probabilities, the so-called sure-thing principle. In his *Foundations of Statistics*, Savage ([1954] 1972, pp. 101–102) addressed the critiques he had received regarding this initial exposition of his research. He conceded that "casual observation shows that real people frequently and flagrantly behave in disaccord with utility theory, and that in fact behavior of this sort is not at all typically considered as abnormal or irrational." However, Savage also claimed that this phenomenon can be observed in all theories that claim to be general, without necessarily leading to their rejection. In particular, Savage had already made it clear in the introductory pages of his volume that his theory was normative rather than descriptive: the theory of choice under uncertainty he proposed was not so much intended to describe real behavior as to provide norms of behavior to be followed.

Nevertheless, Savage himself agreed that, if when faced with a simple choice under controlled conditions one finds that an individual who has "conscientiously

[6] In an interview tracing his academic career, Raiffa recalled when, in the mid-1960s, the course taught by his colleague Schlaifer became mandatory in the Managerial Economics curriculum at Harvard Business School. The fact that the volume he co-authored with Schlaifer was studied by all students was "like an existence theorem: it proved that decision analysis was relevant and teachable to future managers" (interview quoted in Giocoli, 2013). Fourcade and Khurana (2013) reconstruct the spread of mathematical methods in business schools.

studied situations in which the theory seems to lead him astray" does not behave as prescribed, the matter deserves close scrutiny, as this behavior may appear paradoxical. Consequently, Savage was compelled to clarify what had happened when, during a break at the Paris Conference, Allais had presented him with a simple choice problem and he had responded with an answer that contradicted his own theoretical framework. His answer, Savage admitted, was not inconceivable for some real decision-makers but it would be challenging for a theoretician to rationalize.

The problem imagined by Allais (1953), which we quote here from Savage's description of it, is as follows. There are two decisions to be made, each concerning two bets whose monetary prizes and winning probabilities are known. The first decision concerns the choice between:

$$\begin{aligned} &\text{BET 1}: \ \$500{,}000 \text{ with } p = 1 \\ &\text{BET 2}: \ \$2{,}500{,}000 \text{ with } p = 0.1, \ \$500{,}000 \text{ with } p = 0.89 \\ &\quad \text{and zero with } p = 0.01 \end{aligned}$$

The second decision concerns the choice between:

$$\begin{aligned} &\text{BET 3}: \ \$500{,}000 \text{ with } p = 0.11 \text{ and zero with } p = 0.89 \\ &\text{BET 4}: \ \$2{,}500{,}000 \text{ with } p = 0.1 \text{ and zero with } p = 0.9 \end{aligned}$$

As most people actually do, Savage preferred BET 1 in the first decision and BET 4 in the second. This choice indicated that:

$$\begin{aligned} &u(500{,}000) > 0.1u(2{,}500{,}000) + 0.89u(500{,}000) + 0.01u(0), \text{ in the first case} \\ &0.1u(2{,}500{,}000) + 0.9u(0) > 0.11u(500{,}000) + 0.89u(0), \text{ in the second case} \end{aligned}$$

But if one rewrites the two inequalities by adding the utilities with the same payoff, it immediately occurs that the two choices are contradictory:

$$\begin{aligned} &0.11u(500{,}000) > 0.1u(2{,}500{,}000) + 0.01u(0), \text{ in the first case} \\ &0.1u(2{,}500{,}000) + 0.01u(0) > 0.11u(500{,}000), \text{ in the second case} \end{aligned}$$

The choice of BET 1 the first time and BET 4 the second cannot therefore be that of individuals who are maximizing their expected utility.

It must be noted that in Allais's example, the probabilities are known, so the situation under consideration is risky, not uncertain. But the question still concerns the application of the axioms underlying the theory of expected utility. Indeed, if there is no utility function that justifies the observed choices, this means that the axioms underlying the theory are not being respected. The fact that Savage had made choices that contradicted the axioms used to prove his theory was clearly a problem. For this reason, in reporting the episode, Savage made it clear that when he first faced Allais's problem, he had acted impulsively, that is, as a real decision-maker, not as

STATES CHOICES	S1 NUMBER DRAWN: 1	S2 NUMBER DRAWN: 2-11	S3 NUMBER DRAWN: 12-100
BET 1	500,000	500,000	500,000
BET 2	0	2,500,000	500,000
BET 3	500,000	500,000	0
BET 4	0	2,500,000	0

Fig. 6.1 Savage's representation of Allais's problem

a rational one (or as a theorist). But after reconsidering the matter in the light of the theory, thus as a rational decision-maker, he realized that his choice was a "error" that he wanted to correct (Savage, [1954] 1972, p. 103).

In his book, Savage gave a matrix representation of Allais's problem using the acts-states-consequences scheme (Fig. 6.1). The probabilities of winning are represented as possible states of the world: state 1 is the drawing of number 1 in a lottery with one hundred numbers, thus a state that occurs with $p = 0.01$; state 2 is the drawing of a number in the range 2–11, thus $p = 0.1$; state 3 is the drawing of a number in the range 12–100, thus $p = 0.89$. When examined in this way, it is obvious that neither the decision between BET 1 and BET 2, nor that between BET 3 and BET 4 really depends on state 3, because in this state the consequences of the choices are the same. By application of the sure-thing principle, the individual's choices must depend solely on what happens in state 1 and state 2. And since at this point the choice has been reduced to two identical situations, this choice can only be the same: if BET 1 is preferred to BET 2, then BET 3 must be preferred to BET 4 (and vice versa).

Savage therefore concluded that a proper representation of the situation allowed him to see that he had violated an axiom that he considered indispensable to a coherent theory of choice. Faced with the possibility of rejecting the axiom in order to maintain his original choice, he definitely preferred to confirm his confidence in the plausibility of the axiom and to admit that he was initially wrong. His was a violation that could be classified as descriptive in nature, and thus without normative relevance. The point is subtle, but important. Savage stated that, "after thorough deliberation," an individual must make it clear whether or not he accepts the implications of the theory. From a theoretical perspective, the relevant choices are the deliberate ones, not the impulsive ones. However, if an individual, faced with an example where the wrong choice (according to the theory) has a "strong intuitive appeal," intends to maintain preferences that are in conflict with the sure-thing principle, he "must abandon, or modify, the principle; for that kind of discrepancy seems intolerable in a normative theory" (Savage, [1954] 1972, p. 102).

It is worth recalling that Savage considered his theory to be certainly applicable when a decision problem is precisely defined, that is, when the choice situation is a 'small world,' and that his only concern was with its indiscriminate use for 'grand world' decisions. But the example proposed by Allais is clearly a well-circumscribed problem, that is, a small world in which it is possible to test the decision-maker's choices to see whether they conform to the axioms underlying the theory. It was therefore essential for Savage to clarify why an actual choice such as the one he too had made should be corrected. According to Allais (1953, p. 527), the example of choice he proposed to Savage showed the "pseudo-evident character" of the axiom that Savage had presented in his notes for the Paris Conference. But Savage ([1954] 1972, p. 20) believed that he could use his theory "to police my own decisions for consistency": by rewriting the problem in matrix form so that the result of the application of sure-thing principle was self-evident, Savage had ended the dispute.

After Savage's rationalization, the dispute over the sure-thing principle seemed at the time to be over. In private correspondence with Savage, Allais insisted that a relevant psychological aspect could not be accommodated by Savage's application of the axiom: the certainty of winning nothing and the certainty of winning a large sum of money such as $500,000 are not the same thing for a person, so the consequences of state 3 may still have an influence on her choice. As we shall see in the next chapter, the significance of a number of psychological aspects in real-world decision-making was later justified by two Israeli scholars, the psychologist Daniel Kahneman and the mathematical psychologist Amos Tversky, who made extensive use of Allais's examples to provide evidence of descriptive violations of Savage's axioms. However, Allais did not publish the results of his empirical research on the subject until 1979, nor was he able to propose an alternative criterion on an axiomatic basis, so his critique was classified as a paradox. The so-called Allais Paradox was acknowledged as a relevant descriptive violation of the theory, but dismissed as irrelevant as regards its normative implications.[7]

However, in a 1961 paper entitled "Risk, ambiguity and the Savage axioms," the American decision theorist Daniel Ellsberg (1931–2023), then a researcher at the RAND Corporation in Santa Monica, California, proposed a new example of a possible violation of the axioms of choice. Ellsberg's theoretical aim was to test the behavioral relevance of the distinction between risk and uncertainty, which Savage's probabilistic interpretation denied, but which Ellsberg, like Knight, found intuitively relevant. Although a scholar of game theory and an active participant in the ongoing debate, Ellsberg never became a professional academic. He won a prestigious research fellowship at Harvard in 1957, but he was soon attracted by the opportunity to work on studies of the strategic nature of military decisions that the US Army commissioned to the RAND Corporation, a research center visited by many mathematical economists of the time. In this context, he had realized that typical decisions in war scenarios were often characterized by uncertainties that were difficult to categorize as risks, and that, for example, the inability to comprehensively characterize the probabilistic scenario following a possible nuclear attack by an enemy—with specific regard to the reliability of a chain of command in the event of an unprecedented nuclear war—also had to find its way into theoretical

[7] The exchange between Allais and Savage is assessed in Heukelom (2014), while the possible normative content of Allais's approach is highlighted in Mongin (2019).

considerations.[8]However, his theoretical defense of Knight's viewpoint started from a very simple problem. The aim of the 1961 article was not so much to highlight the uncertain characteristics of decision-making contexts in general terms—for the analysis of which, as we shall see in the next section, Ellsberg drew in particular on Keynes's *Treatise on Probability*. Rather, he aimed at finding a direct way to undermine the confidence that his academic colleagues had placed in Savage's approach. Ellsberg's reasoning is crucial to our story because he focused on the possibility that there are choices that are incompatible with the assignment of a subjective probability, rather than a utility function as in Allais. In this sense, Ellsberg's attack can be considered as more fundamental than Allais's.

We already met Ellsberg in Chap. 4 as a critic of von Neumann and Morgenstern's max-min solution. He argued that it is one thing to say that the max-min strategy is a reasonable strategy for a zero-sum game, but quite another to say that it is the only possible reasonable—hence rational—strategy. Likewise, Ellsberg was convinced that the maximization of Savage's subjective expected utility was a reasonable decision criterion, but not the only reasonable—hence rational—one. Unlike what had become accepted in decision theory, he did not consider Savage's criterion to be valid in every situation. In both game theory and decision theory, it was a matter of correctly identifying the kind of uncertainty that pervades the decision context, and then assessing how it should be treated analytically.

Like Allais, Ellsberg sought to identify a small world in which the decisions of individuals could be analyzed in a very simple and direct way, so that a violation of the sure-thing axiom could be easily demonstrated. But unlike Allais, Ellsberg (1961, p. 646) also wanted to identify situations in which the violation of the axiom was deliberate, situations in which "many otherwise reasonable people neither wish nor tend to conform to the Savage postulates." That is, a context of choice that might lead a decision-maker familiar with the theory, in principle Savage himself, not to change behavior contradicting it, even after a thorough reconsideration of the problem. Thus, a situation in which Savage's response to Allais might not appear justified, since even a decision-maker like Savage would admit that there was no mistake to correct.

In seminars at RAND and at various universities, Ellsberg proposed to a number of colleagues two problems in which degrees of belief could be measured in a well-defined way. The first problem involves betting on the drawing of a colored ball from two urns. In a first urn, it is known that there are 100 balls, either red or black, but in an unknown proportion. In a second urn, it is known that there are 100 balls, of

[8] As detailed in Zappia (2021a), in a 1960 document for internal RAND use only, Ellsberg assessed the command and control problems associated with "post-attack capabilities." Ellsberg examined the ability of the chain of command to actually enforce decisions made by commanders in the event of a surprise attack from the enemy. In his advice to the military, Ellsberg strongly emphasized the necessity for a strategy based on flexible, adaptive behavior rather than reliance on what tended to be a series of "prehostilities contingency plans," however detailed they might be. His concern was that the attitude of the military was to neglect that "the general problem is that of dealing with surprise." As a result, he made it clear that the US Army should focus not so much on the occurrence of events considered "unlikely," but on the "possibility of happenings which were considered 'impossible' or were undreamt-of."

which it is also known that 50 are red and 50 are black. A person is asked to bet on the drawing of a red ball or on the drawing of a black ball from the first urn: if the ball extracted coincides with the one bet on, she wins a sum of $100, otherwise she wins nothing. The person is then asked to do the same with the second urn, again for $100. Ellsberg observed that it was normal to receive the same answer in both cases: the typical decision-maker is indifferent between betting on red or on black in front of both the first and the second urn. Thus, following Savage's reasoning, the decision-maker's behavior indicates a subjective probability $p(red_I) = p(black_I) = ½$ and $p(red_{II}) = p(black_{II}) = ½$ (where the suffix indicates the urn). As prescribed by the Ramsey-de Finetti-Savage approach, the choices reveal degrees of belief that can be represented in terms of a probability function.

Next, the same person is asked to bet that a red ball will be drawn from either the first or second urn. Again, if the ball drawn matches the one on which the bet was placed, a sum of $100 is won; otherwise, nothing is won. Ellsberg claimed that the decision-makers he interviewed generally preferred to bet on the red ball drawn from the second urn. He also claimed that the same was true when decision-makers were asked to bet on a black ball being drawn from either the first or second urn: it was common for decision-makers to prefer betting on black being drawn from the second urn. But these choices signal $p(red_{II}) > p(red_I)$ and, simultaneously, $p(black_{II}) > p(black_I)$: the subjective probabilities derived from choices are contradictory.[9]

Although the problem was very well circumscribed, Ellsberg noted, there were many decision-makers whose choices were guided by degrees of belief that did not seem to be representable as ordinary probabilities. Indeed, it seemed entirely realistic to Ellsberg that a person who wanted to bet on, say, red, would prefer to do so from an urn in which there were certainly 50 balls of that color, rather than from one that might contain more, but also fewer. Ellsberg argued that the obvious conclusion was that a decision-maker puzzled by the lack of information about the first urn chose to bet on the drawing from the second.

Ellsberg also proposed a second problem, in which individuals had to bet on the extraction of a colored ball from a single urn. The urn contains 90 balls, of which 30 are red and 60 are either black or yellow, in some unknown proportion. An individual is asked to bet either on the drawing of a red ball or on the drawing of a black ball: if the ball extracted from the urn matches the one bet on, he wins a sum of $100, otherwise he wins nothing. Ellsberg found that it was normal for the decision-makers he tested to respond that they preferred to bet on the drawing of a red ball. The individual is then asked to bet on the drawing of either a red or yellow ball or on the drawing of a black or yellow ball: if the drawn ball coincides with the one that was bet on, a sum of $100 is won, otherwise nothing. Ellsberg noted that in this second case, it was

---

[9] In particular, the elicited probabilities do not sum to 1. If the degrees of belief expressed by the choices were probabilities, it should be true that $p(red_I) + p(black_I) = 1$ and $p(red_{II}) + p(black_{II}) = 1$. But while this is indeed true for the first two choices, when we turn to the second two choices there are two possibilities: either it is true that $p(red_{II}) + p(black_{II}) > p(red_I) + p(black_I) = 1$, or it is true that $p(red_I) + p(black_I) < p(red_{II}) + p(black_{II}) = 1$. The property of additivity of the probability measure is violated.

normal for the decision-makers he tested to respond that they preferred to bet on the draw of an either black or yellow ball.

As in the first problem, even in this second problem, the probabilities that can be derived from the two choices are contradictory: from the first choice one deduces $p(red) > p(black)$, from the second one $p(black\ or\ yellow) > p(red\ or\ yellow)$. But since it can be assumed that $p(red) = 1/3$ and $p(black\ or\ yellow) = 2/3$ this would imply $p(black) + p(red\ or\ yellow) < 1$: thus the subjective probabilities deduced from the choices do not satisfy the standard properties of a probability. Again, it seemed realistic to Ellsberg that a decision-maker would prefer to bet on the fact that, in the first case, there are definitely 30 balls of red color and, in the second case, there are definitely 60 balls of either black or yellow color.

This second problem is important because it allowed Ellsberg to provide a matrix representation in the style of Savage, evidently designed to anticipate the possible objection—that already formulated by Savage in his commentary on Allais—that an adequate representation would exclude the discrepancy with the theory. Anticipating the possible objection, Ellsberg represented the decision problem using the matrix reproduced in Fig. 6.2. If the choice is to bet on red, in the first case, and on either black or yellow, in the second case, the violation of the axiom of the sure-thing becomes apparent because state 3—the drawing of a yellow ball—should not influence the choice. However, Ellsberg noted, most of those who had violated the axiom did not intend to revise their decisions.

In conclusion, in both Ellsberg's examples, it was common to find decision-makers whose choices were deliberately—not impulsively—in violation of the subjective probability theory underlying Savage's version of expected utility theory. Ellsberg's argument was proved using the strategy employed by the scholars whom he identified as "neo-Bernoullians," the scholars inspired by Daniel Bernoulli to generalize the approach to decisions under uncertainty. As in the whole subjectivist tradition that we saw summarized by Raiffa and colleagues in the previous section, Ellsberg ([1962] 2001, p. 132) assumed that it was a matter of discovering the individual's preferences and degrees of belief in the context of "artificially simple choice situations," with the aim of deducing a behavioral style that could then be used to choose between "complex and unfamiliar alternatives." Thus temporarily turning away from the complex problem and constructing simple decision problems from the study of which a method of choice for other decisions can be derived.

STATES ACTS	S_1 DRAWN BALL: RED	S_2 DRAWN BALL: BLACK	S_3 DRAWN BALL: YELLOW
BET ON RED	100	0	0
BET ON BLACK	0	100	0
BET ON EITHER RED OR YELLOW	100	0	100
BET ON EITHER BLACK OR YELLOW	0	100	100

Fig. 6.2 Ellsberg's second problem

In examining his examples as thought experiments, Ellsberg found it intuitively reasonable that a decision-maker might behave in violation of the axiom when faced with an 'ambiguous' urn—the first urn in the first example—or an 'ambiguous' part of an urn—the black or yellow part of the urn in the second example. But he then moved on to test individuals while they were actually making such choices. When presenting these choice problems to colleagues, Ellsberg found that some of them were unwilling to change their own choices even though they contradicted the axiom, thus implicitly showing that they shared his perplexity from a normative point of view.[10] Ellsberg (1961, p. 656) concluded that the issue under discussion was precisely the one that Savage's methodology had highlighted: in the face of a deliberate violation by "a number of people not only sophisticated but reasonable," the theory had to be modified to account for what he called the "ambiguity of information."[11]

6.4 Ambiguity and Uncertainty in Ellsberg

Fifty years after the publication of his article, Ellsberg (2011, pp. 222–223) reconstructed his own state of mind as a young researcher as follows:

> Savage's innovative generalization of the Bernoulli proposition, following Frank Ramsey, was to conceptualize 'subjective probabilities' corresponding to betting odds that people would demand or offer in choosing among gambles: not as 'given' but as derivable for all uncertainties, for a given person, from observing his or her choices. This implied that all uncertainties could be expressed, for all purposes of describing or explaining or predicting or prescribing behavior, in the same probabilistic terms as spins of a well-balanced roulette wheel. Compelling as I did find Savage's axioms and logic, I still found this hard to believe.

[10] As noted above, Ellsberg illustrated his examples during faculty seminars he gave in the late 1950s, and used the presentations to test people in the audience. The people tested during these seminars, "under absolutely non-experimental conditions" (Ellsberg, [1962] 2001, p. 134), were therefore mostly faculty members at various universities and colleagues at the RAND Corporation. But as soon as the methodology of experimental economics began to spread, Ellsberg's results were widely confirmed. Slovic and Tversky (1974) showed that the overwhelming preference to violate Savage's sure-thing principle was confirmed by experimental subjects who had ample opportunity to become acquainted with the principle while tested.

[11] Ellsberg claimed (1961, p. 656) that he also proposed his example to Savage, and that Savage violated his own axiom, without showing any willingness to change his decision after realizing it, thus deliberately. Savage never commented on Ellsberg's work, either in his published writings or in the correspondence available in the archives of Yale University, where his papers are collected. Therefore, it can be argued that he probably considered Raiffa's explanation of the matter, which we will mention below, to be definitive. In private correspondence with de Finetti, though, Savage did not take the opportunity to deny that the violation had been deliberate. When asked by de Finetti whether he had read Ellsberg's account of his behavior, Savage acknowledged to have been in contact with Ellsberg, but simply stated: "I feel that there may be a grain of truth in what he is trying to say, but I find it very difficult to clear my mind on the matter" (excerpt from an exchange between de Finetti and Savage in March 1962, quoted in Zappia, 2021b).

His goal, Ellsberg recollected, was to compel his fellow decision theorists—whom he called neo-Bernoullians—to reconsider the scope of subjective expected utility theory. To do this, he believed that presenting simple problems that showed a flagrant violation of the theory's axioms could have a greater impact than introducing general methodological statements and theoretical arguments about uncertainties that were not risks.

Ellsberg's strategy was to present an example that, according to Karl Popper's logic of scientific inquiry, would falsify the theory; an example which might at first seem only paradoxical, but which would force the proponents of the accepted theory to reflect on the relevance of the violation and ultimately to abandon the theory. In his doctoral thesis at Harvard, written shortly before his definitive involvement as a consultant to the US government, Ellsberg ([1962] 2001, pp. 50–51) recalled that Savage had pointed out to him that this was the strategy used by Bernoulli, when he questioned the criterion of the maximization of the expected value. A widely accepted principle was challenged by its application to a specific problem to such an extent that, as we mentioned in Chap. 2, the mathematician Cramer chose the practical solution of ordinary people, rather than the theoretical solution suggested by his mathematical colleagues of the time. With the St. Petersburg Paradox, Bernoulli had disrupted the dominant theory of the time—the maximization of expected monetary gains—and proposed an alternative theory—the maximization of the expected utility of those gains. And although Bernoulli's theory was not fully elaborated until more than two centuries later, it definitely changed the way in which risky prospects were dealt with.

Ellsberg aimed at a similar result, and indeed he introduced his main investigation of the subject, presented in his doctoral dissertation, by addressing the neo-Bernoullians in a direct way: "Very simply" he claimed "I wish to change their minds" (Ellsberg, [1962] 2001, p. *liii*). But once it was established that his examples were a 'paradox' of the theory of choice—as classified by the statistician Harry Roberts (1963), a colleague of Savage at the University of Chicago, in order to criticize them—what alternatives did Ellsberg have to propose?

Ellsberg's argument was basically put forward in two steps. In his 1961 article, he characterized the decision problem of his examples as a choice under conditions of ambiguity. It was of course evident to Ellsberg that there was no real uncertainty when considering the urns of his examples. In fact, if we go back to the first example, that of the two urns, of the urn whose composition is not known, it is nevertheless known that there can be from 0 to 100 balls of red color and, at the same time, from 100 to 0 balls of black color. The list of possible states of the world is thus exhaustive, and there cannot be the surprise of drawing, for instance, a ball of a different color. There is certainly no uncertainty *à la* Shackle, who believed that the possibility of surprise was an ineradicable feature of economic reality and constituted the most important element of Keynes's discussion of uncertainty.

However, the degree of confidence associated with the information available with respect to the first urn differs from that with respect to the second, to the point of causing subjects to violate the axioms of choice theory. According to Ellsberg, his cases well represented situations in which the available information was perceived as scarce, obscure and vague or accurate but contradictory, as when different experts are consulted and they make predictions that are precise but different from each other. So, in addition to risk and uncertainty, Ellsberg introduced a third situation to consider, which he called 'ambiguity:' "a quality depending on the amount, type, reliability and 'unanimity' of information, and giving rise to one's degree of 'confidence' in an estimate of relative likelihood" (Ellsberg, 1961, p. 657).

Today, Ellsberg's work is largely remembered precisely for introducing this notion. Ambiguity characterizes a decision context that is not simply risky, but neither is it an unspecified uncertainty irreducible to risk. Ambiguity serves to identify those situations in which a less than full degree of confidence in one's degrees of belief manifests itself in the fact that the subjective probabilities inferred from choices are not ordinary probabilities, thus precluding the possibility to employ them to maximize expected utility. In the article, however, the issue was mostly confined to simple problems, which in any case were characterized by information with certain symmetric characteristics, to such an extent that the proponents of Savage's theory had the opportunity to suggest variations in the representation of the specific problem that made it appear, to simplify, less ambiguous than it initially appeared.

In the same issue of the academic journal in which Ellsberg's article was published, there appeared a critical comment by Raiffa, whom we have already identified as one of the main exponents of the new school—a prominent neo-Bernoullian. In his comment, Raiffa (1961) admitted both that he had violated the axioms with his choices when first tested and that he had later verified with his students in class that many of them had done so as well. But his conclusion was that, when he presented them with a different representation of the same problem, the observed results were back in accordance with the theory. Raiffa found that in Ellsberg's first example, the decision-makers could secure a ½ probability of drawing a ball from the first urn, the ambiguous one, if they tossed a coin, before betting, to decide which color to bet on. The introduction of a choice randomization mechanism, Raiffa argued, changed the decision-makers' perception of the first urn, and the different view of the original options made the typical choices compatible with the theory again.

We have already seen this way of arguing above. As in the case of Savage with Allais, the answer to the paradox is entrusted to a representational revision of the problem that had generated it. The new representation would allow the problem to be interpreted by decision-makers in a way that would change their attitude toward the different available options. In response, Ellsberg argued that Raiffa's presentation of his problem substantially changed the nature of the original choice. But even in this case, the debate was essentially over before it had even begun: Savage's theory was a crucial new piece of rational choice theory and proved useful in a wide range of applications, so Ellsberg's criticism remained at best a paradox.[12]

[12] Ellsberg's personal affairs also contributed to this outcome. He did not publish his 1962 dissertation, including the details of his rejoinder to critics like Raiffa, and he did not speak on the subject again after a last brief response to Roberts's criticism in 1963. In fact, he had already become a consultant to the US Department of Defense and was then sent to Vietnam to assess on the ground the progress of the war, declared by the United States in 1964. On his return to the United States, he became skeptical about a possible positive outcome of the US intervention and highly critical of

But Ellsberg did not want to limit his contribution to a circumscribed counterexample. There was a larger theoretical argument on which he based his critique, which he developed in what can be considered a second step in his analysis, the above-mentioned 1962 doctoral dissertation.[13] In his dissertation, Ellsberg presented a much broader view of the issues raised the year before. Specifically, there was a lengthy discussion of decision criteria intended as alternative to expected utility maximization. Ellsberg had already laid the groundwork for analyzing the problem in the 1961 journal article, but now he clarified that even ambiguity could be represented in subjective terms, as long as it was admitted that the decision-makers might not have sufficient confidence in their beliefs to have a single, reliable subjective prior probability.

If the information in their possession was perceived as "limited, unreliable and ambiguous," it could be assumed that decision-makers have not a single, but several subjective probability functions that they think may account for their (lack of) knowledge. It would thus be possible to imagine that they might assign to events probability intervals rather than precise probabilities; or that decision-makers might want to guard against the possibility that the single, "best guess" probability function they had in mind was not reliable, and therefore might want to combine the employment of that single function with other considerations, such as caution against worst-case consequences.

According to Ellsberg, the Bayesian dogma of the existence of a single, precise probability function that can always be associated with the set of events that condition consequences in uncertain environments expressed a narrow view of the way decision-makers are used to represent the external world. He found such a view incompatible with his notion of ambiguity and thus *a fortiori* with uncertainty. His proposal was to start from a set of a priori "plausible" subjective probabilities, rather than from a single, certainly reliable probability as in Savage, and then to examine different decision criteria allowing for such a set of priors. In a perspective that we might call generalized Bayesian, as opposed to Savage's too-strict Bayesian approach—and which Ellsberg ([1962] 2001, p. 165) called "neo-Bayesian" in order to distinguish it from Savage's "neo-Bernoullian"—the criterion might take into account, for example, the breadth of the set of plausible functions; or it might be a weighting among the various plausible probability functions that captures the more or less optimistic attitudes of the decision-maker, as Arrow and Hurwicz had already suggested for the case of complete ignorance. Even the combination of focus values

---

the continuing escalation. When Secretary of Defense Robert McNamara disregarded an extensive series of confidential studies that came to the same conclusion, Ellsberg decided to leak them to the press. Identified as the source of the 1971 publication of the so-called Pentagon Papers in the New York Times, he fortunately escaped prosecution by the court before which he appeared, because his lawyers proved government misconduct in gathering evidence against him. Ellsberg would later devote himself to activism against American expansionism (Ellsberg, 2002).

[13] It was only forty years after the article that had introduced the paradox, after a period of total detachment from economic analysis, that Ellsberg decided to publish the volume entitled *Risk, Ambiguity and Decision,* which reproduced the 1962 doctoral thesis in its entirety (Ellsberg, [1962] 2001).

proposed by Shackle, Ellsberg noted, could find justification in this context. The various solutions that Arrow had initially considered in his 1951 review, but then rejected in favor of the subjective approach in the making, were to be revived.

However, Ellsberg was unable to give an axiomatic foundations for what he was proposing, and this aspect was considered diriment by the mathematical economists he was trying to convince, as we know from both Arrow's review and Koopmans's methodological considerations. Moreover, Savage himself ([1954] 1972, pp. 57–58) had anticipated possible criticism on this front: in his book he had admitted that there did indeed "seem to be some probabilistic relations about which we feel relatively 'sure' as compared with others." But he had then concluded that since he did not know how to formally characterize the issue, he did not think it was useful to go into it in any depth. After a few years, possibly with reference to Ellsberg, Savage (1962, p. 165) reiterated his point: "some people see the vagueness phenomenon as an objection [to precise personal probabilities]; I see it as a truth, sometimes unpleasant but not to be escaped by a new theory."[14]

Apart from a more thorough examination of alternative decision rules, Ellsberg's dissertation is also of great interest because the author took pains to accurately reconstruct the theoretical and methodological context in which his considerations could be placed. There was, again, an explicit reference to Knight. As he had already done in the 1961 article, Ellsberg argued that his examples showed the behavioral significance of the Knightian distinction: it cannot be useful to assume that one can reduce all uncertainties to risks if, given simple examples, the actual choices do not allow one to associate precise probabilities with degrees of belief. But in his search for supporting arguments in the literature on probability theory, Ellsberg found that Keynes's *Treatise on Probability*—which he claimed not to have read until he was working on his doctoral thesis—offered him a much richer interpretative milieu in which to better express his views.

Ellsberg noted that Keynes had devoted an entire chapter to the weight of argument in order to highlight the importance of the confidence an individual has in a probability valuation, with low weight related primarily to "scantiness of evidence of any sort" (Ellsberg, [1962] 2001, p. 11). He then quoted from the chapter of the *Treatise* on the application of probability theory to conduct to note that Keynes wanted to make room for confidence while discussing decision rules: in order to examine how decisions are made in uncertain environments, Keynes ([1921] 1973), p. 345) had argued that "the degree of completeness of the information upon which a probability is based does seem to be relevant, as well as the actual magnitude of the probability, in making practical decisions." Keynes had evidently anticipated, Ellsberg admitted, both the

---

[14] It must be noted, however, that at the urging of another critic of the restricted Bayesian approach, William Fellner, who invited him to speak clearly about Ellsberg's examples, Savage showed a more sympathetic attitude. In a private correspondence reconstructed in Zappia (2021b), Savage argued that he was in principle willing to dispense with the sure-thing axiom, but only on the condition that he be offered an equally plausible alternative axiom on which to base an equally systematic theory of decisions.

notion of ambiguity and the link between the quality of information and decision-making that might render the maximization of expected utility inapplicable.[15]

Moreover, Ellsberg's reconstruction pointed out that the special kind of probability that Keynes had put forward with his logical approach deserved high consideration. He stressed that with his notion of non-numerical probability, Keynes had inspired authors such as the French-born American mathematician Bernard Koopman and the English statistician Irving Good. They both followed in Keynes's footsteps in attempting to develop a logic of degrees of belief based on the rejection of the idea that it is always possible to assign precise numbers to probabilistic judgments, and thus to provide a complete ordering of the probabilities associated with uncertain events.

According to Ellsberg ([1962] 2001, p. 10), the theoretical approach inaugurated by Keynes "admits vagueness as an explicit factor [in decision-making] without apology and provides a formal vocabulary for discussing it." And although these authors' considerations were based on introspection, thus were typical of an 'intuitive' approach to probability, their validity could also be proven on the basis of observed behavior, thus within an 'operational' approach. Ellsberg therefore believed that his examples showed how the behavior observed in controlled contexts—the only really relevant variable for the pragmatists Ramsey and de Finetti and the operationalists de Finetti and Savage—justified Keynes's intuitions, which therefore had to be revalued.

Ellsberg also noted that it was precisely by following Keynes's way of arguing that he could reject the main argument about the consistency of individual degrees of belief that underlies the Bayesian viewpoint expressed by Ramsey, de Finetti and Savage. We have seen in the previous chapter that the Dutch book argument—the consistency condition for degrees of belief that justifies their representation by a probability function—requires that a person's subjective probabilities be derived from a context in which she is obliged to bet. Once the probability has been derived as the maximum price the person is willing to pay to buy a bet, the complement to 1 of that probability should also reveal the minimum price at which she should be willing to sell the same bet, thus as if she were not the bettor but the banker.

Ellsberg ([1962] 2001, pp. 68–70) observed that the pioneering works of de Finetti and Ramsey, although essential to the development of decision theory, did not provide a proper axiomatic justification for the coincidence between the maximum price, in the first case, and the minimum price, in the second case. Their argument was simply that it must be rational to do so in order to avoid sure losses. However, Ellsberg added, one of the few English statisticians of the time who advocated the subjective approach to ground probability, Cedric Smith, found the non-coincidence of the two prices to be perfectly reasonable and endeavored to develop a Bayesian statistical theory of decisions on this basis.

---

[15] In fact, as we saw in Chap. 3, Keynes also anticipated the example of the two urns. Ellsberg deserves credit for studying the case of the urns from a behavioral point of view, that is, for reasoning about subjective probabilistic evaluations elicited by observed choices. But Keynes's intuitive argument, that there is a different degree of belief in evaluating the draw from one urn rather than from the other, completely anticipates Ellsberg's.

That the probabilities do not add up to 1—the feature that is revealed by the paradoxical choices in Ellsberg's examples—was regarded by Smith (1961) as a perfectly obvious possibility, even when the probabilities are derived from the willingness to bet, not just in Keynes-like intuitive terms. Ellsberg referred to an early work by Smith, but for the sake of clarity we quote here from a later paper in which Smith (1965, p. 478) gave the following example: "if I am willing to BET 2 to 1 on sun against rain, and 1 to 4 on rain against sun, this means that I regard sun as between 2 and 4 times as probable as rain; and I do not need to be more precise than this." Smith simply observed that, in order to determine the subjective probabilities, two independent measurement for each event were necessary, one while betting on its occurrence and the other while betting against it. Proceeding this way, it was not unusual to find that the two probabilities do not add up to 1.[16]

Smith concluded that, although they were derived from observations of behavior, as prescribed by the Bayesian approach, "probabilities and utilities are no longer uniquely defined, but, in accordance with human vagueness and imprecision, they are only determined within a certain range." For Ellsberg ([1962] 2001, pp. 68–69), it was therefore possible to argue that the 'imprecise' Keynes-Koopman-Good intuitive approach to probability had found its operational version in Smith, where the Ramsey-de Finetti-Savage dogma of a single numerically precise prior probability was removed. A system of degrees of belief represented in terms of interval-valued probabilities would not determine a complete ordering of events, since intervals may overlap. Yet "a partial ordering of events … in terms of probability would still be possible" (Ellsberg, [1962] 2001, p. 73), vindicating Keynes's intuition that it must be admitted that, sometimes, a person might not be ready to bet in favor of against an event.[17]

In conclusion, in *Risk, Ambiguity and Decision*, Ellsberg proposed a series of theoretical arguments that, although all related to the urn examples, presented a generalized subjective approach to probability, intended as alternative to the neo-Bernoullian approach. He did not simply reject Savage's conclusion that all uncertainties are representable as risks. In his doctoral thesis, it emerged that the main critical argument against Savage's theory of subjective expected utility, the so-called

---

[16] The probabilities derived from the two separate bets are $p(sun) = 2/3$ and $p(rain) = 1/5$, so that $p(sun) + p(rain) = 13/15 < 1$: the elicited probabilities do not satisfy the additivity property. But it is not true that the individual would certainly lose by showing, as de Finetti would say, such an inconsistency. If he is really allowed to bet on what he wants—as is usually the case—he will find it convenient not to bet on either sun or rain for prices associated with the interval $2/3 < p(sun) < 4/5$, i.e., $1/5 < p(rain) < 1/3$. It can also be noted that if the subject is allowed to bet only on sun, betting at 2/3 reveals that he considers sun more likely than rain twice; but if he is allowed to bet only on rain, betting at 1/5 reveals that he considers sun more likely than rain four times. Thus, the bet results in a range of probabilities, for sun from 2/3 to 4/5 and for rain from 1/5 to 1/3. The 'imprecise' probabilities are interval-valued probabilities.

[17] Runde (2001) and Feduzi et al. (2014) assess this aspect of the Keynesian-oriented criticism of the subjective approach, including de Finetti and Savage's perplexed reaction to Smith's argument. Levi (1980) outlines a philosophy of decision-making justifying the use of 'imprecise' probabilities, while Walley (1991) develops a comprehensive approach to the statistical applications based on them.

Ellsberg Paradox, was much more than a paradox because it stemmed from a genuine philosophy of probabilistic judgement for uncertainty. Moreover, Ellsberg made it clear that this philosophy of decision-making was largely anticipated by Keynes. However, personal events kept Ellsberg away from the debate, and the dissertation was only read at the time by Savage, de Finetti and a few other scholars, almost all of whom agreed to defend the dominant approach.[18]

6.5 Concluding Remarks

It is difficult to overstate the importance of Ellsberg's contribution to the development of rational choice theory. Initially, his arguments had minimal impact on the evolution of the theory. However, the extensive documentation of violations led to the emergence of numerous theoretical studies on the subject (Camerer & Weber, 1992). Since the mid-1980s, much of decision theory under uncertainty theory has been decision theory under ambiguity, with analytical developments usually justified by an analysis of the significance of the Ellsberg Paradox (Kelsey & Quiggin, 1992). Any alternative decision criterion to subjective expected utility theory has been presented primarily for its ability to explain the behavior of subjects who violate the sure-thing axiom in the ambiguous contexts highlighted by Ellsberg. In addition, some of Ellsberg's intuitions regarding the criteria that would satisfy his idea of generalized rationality in the style of Keynes—versus strict rationality *à la* Savage—have received axiomatic justification, as we will document in the next chapter.

However, it is also to Ellsberg's credit that he identified the methodological context within which the perplexities concerning Savage's solution could be clarified. As Ellsberg noted in commenting on Marschak's view, which was later adopted by Savage, a theory of decisions under uncertainty must be able to exhibit coherence not only "internally" but also "externally." According to Ellsberg ([1962] 2001, p. 28), it is necessary to distinguish the "meaningful, and respectable aspiration to 'prolong' logic and arithmetic into the realm of decision"—which was Marschak's mantra—from the "demonstrated achievement of that goal." In the context of extending the scope of logic to decision under uncertainty, internal consistency alone is inadequate as a criterion. Forcing the behavior of individuals to be consistent with a particular set of axioms necessitates a commitment to making certain decisions that, instead, "must depend ultimately upon individual judgement … not upon traditional logic."

[18] An example of this defense is provided by the attitude of Jacob Marschak, whom Ellsberg identified as one of the colleagues who had violated the theory when he submitted his examples to him. In an article on the possibility of testing the axioms of choice in experimental situations under uncertainty, Marschak (1964, p. 105) admitted that "under stress, or when the problem is complex, or has not been pre-structured for S[ubjects]' use" violations of the axioms might occur. But he then concluded that this was precisely why it was becoming common practice even among managers of large corporations to present to their boards of directors problems that had been pre-structured in a "tabular presentation," a practice which allowed the rationality of the axioms to be applied to real choices.

In essence, Ellsberg's conclusion was that a logical system that satisfies the internal consistency or "aesthetic considerations of form, style or semantic usage" as defined by the scientific community, yet does not align with the "deliberate decisions" of a decision-maker, will induce decision-makers to choose to satisfy their preferences and simply let the axioms remain inapplicable (Ellsberg, [1962] 2001, p. 29).

References

Allais, M. (1953). Le comportement de l'homme rationnel devant le risque: Critique des postulats et axioms de l'école américaine. *Econometrica, 21*, 503–546.

Arrow, K. J. (1951). Alternative approaches to the theory of choice in risk-taking situations. *Econometrica, 19*, 404–437.

Arrow, K. J., & Debreu, G. (1954). Existence of an equilibrium for a competitive economy. *Econometrica, 22*, 265–290.

Blaug, M. (1997). *Economic theory in retrospect* (5th ed.). Cambridge University Press.

Boumans, M., & Davis, J. B. (2010). *Economic methodology: Understanding economics as a science*. Palgrave Macmillan.

Camerer, C. F., & Weber, M. (1992). Recent developments in modelling preferences: Uncertainty and ambiguity. *Journal of Risk and Uncertainty, 5*, 325–370.

Debreu, G. (1959). *Theory of value: An axiomatic analysis of economic equilibrium*. Yale University Press.

Ellsberg, D. (1961). Risk, ambiguity, and the Savage axioms. *Quarterly Journal of Economics, 75*, 643–669.

Ellsberg, D. ([1962] 2001). *Risk, ambiguity and decision*. Garland Publishing.

Ellsberg, D. (2002). *Secrets: A memoir of Vietnam and the Pentagon papers*. Viking.

Ellsberg, D. (2011). Notes on the origins of the Ellsberg urns. *Economic Theory, 48*, 221–227.

Feduzi, A., Runde, J., & Zappia, C. (2014). De Finetti on uncertainty. *Cambridge Journal of Economics, 38*, 1–21.

Fourcade, M., & Khurana, R. (2013). From social control to financial economics: The linked ecologies of economics and business in twentieth century America. *Theory and Society, 42*, 121–159.

Friedman, M. (1953). The methodology of positive economics. In Id. *Essays in positive economics*. University of Chicago Press.

Giocoli, N. (2013). From Wald to savage: *Homo economicus* becomes a Bayesian statistician. *Journal of the History of the Behavioral Sciences, 49*, 63–95.

Heukelom, F. (2014). *Behavioral economics: A history*. Cambridge University Press.

Kelsey, D., & Quiggin, J. (1992). Theories of choice under ignorance and uncertainty. *Journal of Economic Surveys, 6*, 133–153.

Keynes, J. M. ([1921] 1973). *A treatise on probability. The collected writings of John Maynard Keynes* (Vol. VIII). Macmillan.

Koopmans, T. J. (1947). Measurement without theory. *Review of Economics and Statistics, 29*, 161–172.

Koopmans, T. J. (1957). *Three essays on the state of economic science*. McGraw-Hill.

Levi, I. (1980). *The enterprise of knowledge*. MIT Press.

Marschak, J. (1964). Actual versus consistent decision behavior. *Behavioral Science, 9*, 103–110.

Mongin, P. (2019). The Allais paradox: What it became, what it really was, what it now suggests to us. *Economics and Philosophy, 35*, 423–459.

Pratt, J. W., Raiffa, H., & Schlaifer, R. (1964). The foundations of decision under uncertainty: An elementary exposition. *Journal of the American Statistical Association, 59*, 353–375.

Raiffa, H. (1961). Risk, ambiguity, and the Savage axioms: A comment. *Quarterly Journal of Economics, 75*, 690–694.

Raiffa, H., & Schlaifer, R. (1961). *Applied statistical decision theory*. Harvard University Graduate School of Business Administration.

Roberts, H. (1963). Risk, ambiguity, and the Savage axioms: A comment. *Quarterly Journal of Economics, 77*, 327–336.

Runde, J. (2001). Chances and choices: A note on probability and beliefs in economic theory. In U. Maki (Ed.), *The economic world view: Studies in the ontology of economics*. Cambridge University Press.

Savage, L. J. ([1954] 1972). *The foundations of statistics*. Dover.

Savage, L. J. (1962). Subjective probabilities and statistical practice. In G. A. Barnard, & D. R. Cox (Eds.), *The foundations of statistical inference: A discussion*. Methuen.

Shackle, G. L. S. (1949). *Expectation in economics*. Cambridge University Press.

Smith, C. A. B. (1961). Consistency in statistical inference. *Journal of the Royal Statistical Society, Series B, 23*, 1–25.

Smith, C. A. B. (1965). Personal probabilities and statistical analysis. *Journal of the Royal Statistical Society, Series A, 128*, 469–499.

Slovic, P., & Tversky, A. (1974). Who accepts Savage's axiom? *Behavioral Science, 19*, 368–373.

Wald, A. (1945). Statistical decision functions which minimize the maximum risk. *Annals of Mathematics, 46*, 265–280.

Walley, P. (1991). *Statistical reasoning with imprecise probabilities*. Chapman & Hall.

Zappia, C. (2014). Non-Bayesian decision theory ahead of its time: The case of G. L. S. Shackle. *Cambridge Journal of Economics, 38*, 1133–1154.

Zappia, C. (2021a). From Knightian to Keynesian uncertainty: Contextualizing Ellsberg's ambiguity. *Cambridge Journal of Economics, 45*, 1027–1046.

Zappia, C. (2021b). Leonard Savage, the Ellsberg Paradox and the debate on subjective probabilities: Evidence from the archives. *Journal of the History of Economic Thought, 43*, 169–192.

Chapter 7
Modern Uncertainty Theories and the Return to Keynes

7.1 Introduction

The theory of rational choice under uncertainty that consolidated after the 1950s rests on four pillars: (i) it assumes the existence of a 'state space' sufficiently elaborate to represent every possible event that could affect the consequences of choices, that is, a space containing every possible piece of information relevant to the outcomes, with a 'state' resolving all relevant uncertainty; (ii) for any event for which no objectively given probability of occurrence is known, it assumes that the decision-maker's opinions are nevertheless represented by a probability, and that there is a single 'subjective a priori probability' function defined over the state space; (iii) each new piece of information is added to the decision-maker's knowledge according to Bayes' rule, on the basis of which a 'subjective a posteriori probability' is arrived at; and (iv) the decision is made by choosing the action that maximizes the expected utility, calculated as the average of the probability-weighted utilities of the consequences.

The economic theory of rational choice is called 'Bayesian' with particular reference to the second of these conditions. Indeed, the validity of Bayes's rule for updating probabilities is unanimously accepted within the probabilistic framework axiomatized by Andrej Kolmogorov, as it follows directly from the definition of conditional probability. What distinguishes the Bayesians from other scholars is their argument that the subjective probability priors to which Bayes's rule can be applied exist in every situation. Bayesian decision-makers do not distinguish between risk and uncertainty because they are always able to assign a precise prior probability to each possible event, regardless of the decision context (Gilboa et al., 2008).

As we have already noted in the previous chapters, there have always been reservations about this structure of argumentation in the debate among economists. For the sake of simplicity, these objections can be traced back to the four pillars just mentioned. For example, even before Leonard Savage's systematization, Maurice Allais had pointed out that the fourth condition was inapplicable, even in the case of objective probabilities, since in some choice situations the observed choices were

C. Zappia, *Uncertainty in Economics*, Springer Studies in the History of Economic Thought, https://doi.org/10.1007/978-3-031-91221-4_7

inconsistent with the maximization of an expected utility function. The fact that the first condition was exceedingly stringent had led Savage himself to state that it was important to distinguish a small world, in which the idea that the decision-maker has a sufficiently elaborated state space could be justified, from the large world of decisions of all kinds. For a staunch critic like George Shackle, the first condition was clearly unrealistic, and therefore to be rejected in any situation of uncertainty in favor of a non-probabilistic representation of beliefs. Daniel Ellsberg, on the other hand, focused on the difficulty of justifying the second condition, arguing that reducing all uncertainties to risks through the use of a single subjective probability function would prove inapplicable even in simple decision contexts that might appear 'ambiguous' even to experienced decision-makers. Ellsberg had taken up ideas already formulated by Frank Knight and John Maynard Keynes and suggested that they should be re-evaluated not only in theory, but also for their behavioral significance: a limited confidence in a probability assessment can be associated to ambiguity aversion.

For a long time, however, the response of the majority of economists to criticisms and paradoxes was one of closure. For example, Jack Hirshleifer and John Riley, among the leading scholars of the economics of information that developed in the 1980s as a major application of subjective expected utility, denied that the "dissident literature" was convincing. They claimed that experiments like Ellsberg's did not imply a rejection of the standard assumption of rationality employed by economists, but rather "merely illustrate certain limitations of the human mind as a computer." It is not uncommon, Hirshleifer and Riley claimed, for the brain to be deceived by the way a question is posed, just as an optical illusion can deceive the eye. However, "discovering and classifying these mental illusions are fruitful activities for psychologists, but these paradoxes are of relatively little significance for economists" (Hirshleifer & Riley, 1992, p. 34).

The defense of the dominant theory has been so vigorous that some of the critics have characterized their own work as only descriptively valid, without really questioning the normative significance of the theory they were criticizing. In proposing their 'prospect theory,' Daniel Kahneman and Amos Tversky intended to explain the behavior of real subjects in situations such as those indicated by Allais; but they did not intend to offer norms of behavior as an alternative to the rule of expected utility maximization, which they continued to regard as valid for prescriptive purposes. The hypotheses formulated by Kahneman and Tversky (1979) to describe observed behavior obviously concern both probability and utility, but with some significant transformations. For probabilities, they considered them to be more akin to 'decision weights,' because in their empirical studies they observed that it is typical for real decision-makers to overestimate low probabilities and underestimate high probabilities, signaling that decision weights do not satisfy the additivity property of a standard probability function. For utility, they found that the same amount of money is valued differently if it is a gain (an increase in wealth relative to an initial situation) or a loss (a decrease in wealth relative to an initial situation), an attitude indicative of 'loss aversion,' which implies a discontinuity in the shape of the utility function.

Kahneman and Tversky concluded that individuals' actual choices can be represented by a weighted average of a utility function exhibiting loss aversion with decision weights that do not satisfy the properties of probabilities. On this basis, a large literature developed, both on the experimental side—highlighting a variety of violations of individual rationality—and on the theoretical side—aiming to find axiomatic foundations for behaviorally motivated rules of behavior (Wakker, 2010). But the goal of prospect theory was not to replace expected utility theory, because it was still believed that decision-makers made mistakes when they violated the axioms of expected utility: preference anomalies such as "inconsistencies, intransitivities and violations of dominance" lead to "normatively unacceptable consequences" that may exist only because the decision-maker "does not have the opportunity to discover that his preferences could violate decision rules that he wishes to obey" (Kahneman & Tversky, 1979, p. 277).

However, in the face of mounting evidence of how pervasive and systematic violations were, under conditions of both risk and uncertainty, attitudes were bound to change, even among decision theorists. While Hirshleifer and Riley reiterated the necessity for unquestioning adherence to the dominant paradigm, the decision theorist Mark Machina admitted that what he himself had regarded until a few years earlier to be "one of the 'success stories' of economic analysis"—both because of its solid axiomatic foundations and because of the impetus it had given to the "newly emerging information revolution in economics"—now appeared to him to be a field of study in which "unsolved problems" were at least as important as solved ones. The theory of choice under uncertainty seemed to him "a field in flux" because of the growing empirical challenges and their normative implications (Machina, 1987, p. 121).

The approach proposed by Machina followed the lines already seen in the previous chapters: first, he illustrated some of the problems that have challenged the dominant model; then, he presented the empirical evidence that highlights the importance of violations; finally, he explored the necessary change "in the way we view and model economic behavior under uncertainty" (Machina, 1987, p. 122). In the 1980s, 'non-expected' utility theories for explaining choice situations under risk, and 'non-additive' subjective probability theories for the representing situations of uncertainty, began to make their way into research.

This new field of discussion includes a group of scholars who, with explicit reference to Keynes's teachings, have always rejected the reduction of uncertainty to risk. Paul Davidson, a leading exponent of the so-called Post Keynesians, commented on Machina's critical review by noting that while it raised questions about the actual relevance of the theory of subjective probability, its author did not discuss the possibility that there might be alternative ways of understanding uncertainty, that is, models that exclude that beliefs can always be represented by a probability measure. Rather, the post Keynesian position was (and still is) that "there are many important situations where 'true' uncertainty exists regarding future consequences of today's choices," and that "whenever conditions of true uncertainty prevail, human behavior may differ systematically from what is implied by the standard expected utility perspective" (Davidson, 1991, p. 130). To illustrate what he meant by "true" uncertainty, Davidson recalled the Keynesian quote that has punctuated our history since the introductory chapter of this volume, stating that truly uncertain situations are those in which "there is no scientific basis on which to form any calculable probability," situations about which—as Keynes (1937, p. 214) had argued—"we simply do not know."

This concluding chapter proceeds in two directions. First, it attempts to outline the ways in which the evidence related to Allais's and, in particular, Ellsberg's paradoxes actually oriented research in decision theory in the last part of the twentieth century. Second, the post Keynesian perspective is presented and an attempt is undertaken to assess whether Davidson's rejection of any probabilistic approach constitutes the sole possible reinterpretation of Keynes's contribution. The chapter then explores the broader rationale for what has been termed the return to Keynes, which can be traced back to the financial crisis of 2007–2008.

7.2 Probability and Uncertainty in Modern Decision Theory

We saw in the previous chapter that important violations of Savage's crucial axiom—the so-called sure-thing principle, heir to von Neumann and Morgenstern's axiom of independence—had already been identified in the years when the theory of subjective expected utility was being developed. But the immediate responses of Savage to Allais and Raiffa to Ellsberg essentially ended the debate about the significance of the theory for decades. In his aforementioned 1987 review of the state of decision theory, Machina concluded by conceding that expected utility maximization remained the essential decision criterion because it had been fruitfully applied to the most diverse areas of information economics—search, investment, bargaining and auction models—so much so that any possible alternative could only be incorporated into economic theory and practice if it could be equally broadly applied.[1]

Indeed, the alternatives to expected utility that were developed in the years following Savage's systematization failed to provide similarly effective normative criteria. For instance, an alternative philosophy of choice, such as that proposed by Herbert Simon (1916–2001), never found real application in the theoretical developments of the period. In the early 1950s, Simon was a member of the Cowles Commission, but later broke with it in reaction to what he saw as an overly radical turn toward the axiomatization of economic theory. He acknowledged that the new

[1] We have not in our history insisted on informational issues, which are in many ways a different way of looking at the same problems we classified as uncertainty issues. Information intended as a commodity, as an economic object to be analyzed as part of the decision-making process, was an outgrowth of the studies that had been put forward at the Cowles Commission since the 1950s by many of the same scholars met in the previous chapter, such as Kenneth Arrow, Leonid Hurwicz, Tjalling Koopmans and Jacob Marschak, and others to be mentioned in this chapter, such as Herbert Simon. But the intent of these studies was to discuss issues of knowledge diffusion in an economic system that was no longer visualized under the assumption of perfect foresight. And even at the individual level, the emphasis was not so much on the criteria of optimal decision-making as on the technology of the processes by which decisions were made. For a detailed history of information economics, see Mirowski and Nik-Khah (2017).

axiomatic foundations of utility theory had provided a formal framework for statistical decision theory, from which normative models of management science and operations research had emerged for application to real decision-making problems. But even with these "powerful new tools and machines, most real-life choices still lie beyond the reach of maximizing techniques – unless the situations are heroically simplified by drastic approximations." He therefore concluded that "rationality in real life must involve something simpler than maximization of utility or profit" (Simon, 1959, pp. 259–260).

After moving to the Carnegie Institute of Technology in Pittsburgh, Simon pursued a multifaceted approach to modeling decision-making, focusing in particular on two aspects that he believed severely constrained the actions of real decision-makers: limited informational knowledge and limited computational capabilities. He formulated the hypothesis that a real decision-maker should be viewed as a "*satisficing* animal whose problem solving is based on search activity to meet certain aspiration levels rather than a *maximizing* animal whose problem solving involves finding the best alternatives in terms of specified criteria" (Simon, 1959, p. 278). The so-called bounded rationality approach—according to which, since choices cannot aspire to be optimal, it is necessary to focus on options that are satisfying given the constraints—was initially shared by a group of colleagues interested in the managerial aspects of business and public administration, a group that had formed around Carnegie's Graduate School of Industrial Administration. Focusing on the cognitive aspect of decision-making, in order to characterize the behavior of a real decision-maker in the face of uncertainty Simon sought to identify mechanisms for reducing its complexity.[2]

However, the deviations from rational behavior that he emphasized were soon classified by critics of this approach as merely additional constraints on the behavior of decision-makers, to be analyzed like any other constraint in an optimization problem, simply by extending the set of constraints that were usually thought to influence individual decisions. The Carnegie Graduate School itself gradually became a bastion of the Bayesian approach to management problems, in the style of those at Harvard and Chicago to which we referred in the previous chapter. The normative value of the mainstream view was so widely accepted that—although it spawned a substantial stream of research that led to the eventual recognition of the relevance of his intuition, which came with the award of the Nobel Prize in Economics in 1978—Simon's real influence on the theory of choice under uncertainty remained for a long time confined to applied studies. Eventually, Simon's approach was identified as characteristic of an 'old' behavioral economics, largely subsumed within the 'new' of Kahneman and Tversky.[3]

---

[2] The cognitive component of Simon's work, particularly the possibility that satisficing behavior can generate a kind of path-dependency in the individual's representation of the environment, thereby precluding the interpretation of 'irrational' choices as errors to be corrected, is evidenced in Rizzello (2002).

[3] While some classical assumptions are relaxed in such models, it has been argued that the fundamental methodology has remained unaltered and that behavioral economics has not produced a paradigm shift. Many subsequent models have employed the term 'bounded rationality,' but

Yet, critical considerations such as those proposed by Ellsberg were so important that they have reappeared in analyses in recent years. Itzhak Gilboa (2009, p. 136)—whom we have already identified as one of the most important contemporary scholars of the theory of choice under uncertainty—has argued that there has always been a sense of "cognitive unease" among decision theorists about the Bayesian dogma of a single subjective prior probability, and more generally about the idea that the employment of subjective probability functions can be justified simply by demonstrating their internal consistency. Gilboa reports that David Schmeidler, the Israeli mathematical economist with whom he pioneered the research on non-Bayesian decision theory, commenced his influential work by noting that he saw nothing irrational in an individual preferring to settle issues by tossing a coin known to be fair—for instance, because it had been tested before and found to have a relative frequency of heads of about 50%—rather than by tossing a coin about which nothing is known. The intuition underlying this argument, that the quality of available information must be taken into account, is clearly reminiscent of both Keynes's comment on the two urns we reported in Chap. 3 and Ellsberg's reiteration of his argument that we detailed in the previous chapter.

In the face of systematic violations, Gilboa contends, it is necessary for a theory—even one conceived to provide a norm of behavior and not necessarily an explanation of actual behavior—to be evaluated for its ability to provide guidance that will be found convincing by the individuals who actually make the decisions. We recall from Chap. 3 that underlying Knight's distinction between risk and uncertainty is the idea that there are more or less complex probabilistic contexts. We also noted that this theme has remained ever present in subsequent analyses, both by authors who thought they could deal with it by always equating uncertainty with risk, and by authors who did not consider the reduction of uncertainty to risk to be a general rule. It is from Gilboa that we have taken up the classification of probabilistic contexts in increasing order of conceptual complexity.

Gilboa's point is that when it comes to situations with a probabilistic context characterized by the highest complexity—for example, deciding whether to invest in new fixed capital in the face of the possibility of a financial collapse—a theory of uncertainty must account for the possibility that the decision-maker does not know how to assign a definite prior probability to events:

> wars and financial crises may or may not occur … [and] there is a lot of evidence one may rely on in assessing their probabilities, but coming up with a single probability measure over the state space is typically a challenge. (Gilboa, 2025, p. 30)

Therefore, even if one considers it necessary to base economics on a theory of rational choice—as all of the authors we have focused on in this history have

---

have used it to introduce relatively minor modifications, not to develop models of a completely different nature. Indeed, Simon's satisficing behavior has rarely been used as the basis for behavioral economics. That the behavioral heuristic emphasized by Simon implies a proper normative theory, an alternative to expected utility maximization, has been argued instead by Gigerenzer (2008). Petracca (2021) has reconstructed the steps that led to the normalization of Simon's approach, underlying its inappropriateness.

argued—one still has to ask what kind of rationality one wants to assume under conditions of uncertainty, and therefore one has to reckon with what the individuals observed in making choices find persuasive or not. From the perspective of modern decision theory, as represented by Gilboa and his co-authors Andrew Postlewaite and Schmeidler, it is admitted that—regardless of the plausibility of the axioms suggesting rational action, or the authority of the theorist proposing them—a criterion of behavior can be considered "subjectively rational" for a person only if she can be convinced that a decision of hers that contradicts the criterion is wrong. That is, if she feels so embarrassed by her actual choice that she would want to change it in light of the theory. Similarly, a mode of behavior can only be classified as irrational by a person if it happens that "when the latter is exposed to the analysis of her choices, she would have liked to change her decision" (Gilboa, 2015, p. 316). The affinity between Gilboa's position and Ellsberg's philosophy in discussing Savage's theory, which we illustrated in the previous chapter, is striking (Zappia, 2018).

The most important development in decision theory at the end of the twentieth century—known as Choquet Expected Utility theory to signal both the derivation and the differences from Savage's Subjective Expected Utility theory—stems from the idea of explaining the violations pointed out by Ellsberg by adopting subjective probability functions that do not respect the property of additivity, hence called 'capacities.' Non-additive probabilities are then used as weights to ponder utilities, using an integral developed by the French mathematician Gustave Choquet specifically for these particular probabilities. The axiomatic structure justifying the maximization of Choquet's expected utility is based on an axiom intended to replace the axiom of the sure-thing in order to distinguish, in the sphere of the choice of acts, those that meet Savage's stringent requirements from those that do not, in the sense of Ellsberg (Schmeidler, 1989).

Moreover, as Ellsberg had already pointed out, the fact that probabilities do not sum to 1 can also be explained by assuming that in ambiguous situations individuals assign to events probability intervals, rather than precise probabilities.[4] Given the arbitrariness of assigning a precise probability in a situation where the information

[4] Faced with an ambiguous urn such as the first in Ellsberg's example with two urns, with his choice to bet on the draw of red (black) ball from the second urn rather than the first, ambiguous urn, a subject reveals a subjective probability $p(red_I) < p(red_{II}) = ½$ (and simultaneously $p(black_I) < p(black_{II}) = ½$). This choice can be thought of as that of a decision-maker who reasons by making the assumption that there might be, say, between 40 and 60 balls of both colors in the ambiguous urn, and then finds it reasonable to concentrate on the lowest value. In fact, when placed in front of the first urn taken individually, having no information about the proportion of balls, the decision-maker had assigned probability $p(red_I) = p(black_I) = ½$; but when considering the second urn as well, the ambiguity of the first emerges, and the subject, perplexed by the lack of precise information, may show ambiguity aversion, thus assigning to $p(red_I)$ (and $p(black_I)$) a probability value of, to follow the example, 4/10. Recall that the subjective probability $p(red_I)$ (and $p(black_I)$) can be deduced from the choices made by proposing prices to participate in the bet, so when the subject reveals that it is worth $p(red_I) < p(red_{II}) = ½$, it is possible to assess the subjective probability of $p(red_I)$, and then do the same with the subjective probability of $p(black_I)$. But if it is true for both that the probability is 4/10 then the decision-maker is actually assuming that there may be between 40 and 60 balls of each color.

is ambiguous, it can be assumed that decision-makers' beliefs are represented by a set of subjective probabilities, so that the subjective priors are indeed multiple priors. As a result, the decision-makers' attitude toward the ambiguous situation is revealed by the dimension of this set of probabilities, with complete uncertainty (ignorance) identified as the situation in which they include all the possible probabilities in this set. Gilboa and Schmeidler (1989) suggested that, in this case, Savage's model can be generalized to represent the aversion to choose when facing ambiguous situations. The decision-maker is now supposed to evaluate the expected utility of each act relative to each probability function in the set of those contemplated as possible a priori—called 'plausible' by Ellsberg—and then to compare the lowest expected utilities, one for each act, and choose the act that provides the highest utility among the lowest.

Gilboa and Schmeidler proposed an axiomatization of this possible decision criterion, intended as an alternative to expected utility maximization. Their claim was that in this way a 'rational' choice rule suitable for a decision-maker who manifests 'ambiguity aversion' is well defined. By following this approach—called Max-min Expected Utility—the typical choices in Ellsberg's examples can be justified. Furthermore, it can be shown that the choice criterion that emerges from the first of these two settings—Choquet Expected Utility—selects a choice set that is the same as that obtained by using max-min expected utility, when the probability interval is derived from the maximum price that subjects are willing to pay (as just exemplified in the previous footnote).

The importance of this way of dealing with ambiguity is demonstrated by a number of applications that have been proposed using the multiple probability approach. In particular, it justifies the possibility, denied by the strictly Bayesian approach, that decision-makers decide not to decide, that is, that they are unwilling to change their initial endowment in some situations. Consider an investor in a financial market who is considering whether to buy more shares of an asset she already has in her portfolio, to sell the shares she already has, or to do nothing. If the investor's views about the future stock price are represented by a single, precise prior probability, then she has a single reservation price at which she neither buys nor sells. If the market price is higher than her reservation price, she would sell; if it is lower, she would buy. However, this is not the typical behavior of traders in the financial markets. Apart from considerations linked to transaction costs and other market frictions, traders usually have in mind a maximum market price above which they would sell, a minimum price below which they would buy, and an interval between the two prices within which the portfolio is not changed. Dow and da Costa Werlang (1992) showed that this is exactly what can happen when an investor's subjective probabilities are represented using non-additive probabilities reflecting ambiguity aversion.

But the intuitive content of the multiple probabilities approach has also been considered relevant in macroeconomic analysis. Nobel laureates Lars Peter Hansen and Thomas Sargent (2001) have studied the problem of a policymaker who, when formulating assumptions about the actions to be taken, considers the question of whether the model he is using to represent the economic system is correct or not. The robustness, in the statistical sense, of an economic policy intervention is studied under the assumption that the probabilistic component of the model may not be represented by a single prior, but may in fact be a set of priors. The policymakers' choices can thus be represented in a way that takes into account the confidence they have in the model of the system they are using.

Both of these approaches—non-additive probabilities and multiple probabilities—are elaborated through axiomatic structures that can be used to explain violations of Savage's axioms. Both assume that a narrow Bayesian perspective is inadequate for representing situations in which information is perceived as ambiguous as postulated by Ellsberg, and both derive a decision criterion that fits decision-makers who exhibit an aversion to the supposedly rational choices they would make in contexts perceived as unambiguous. This development marks a significant advance in the direction of decision models that extend the applicability of the theory to more realistic scenarios. Indeed, the formal developments of Gilboa and Schmeidler's approach have spawned an entire literature on decision theory under ambiguity, suggesting increasingly sophisticated decision criteria.[5] However, the question remains whether this account is suitable not only for the ambiguous contexts identified by Ellsberg, but also for 'true' Keynesian uncertainty.

7.3 Probability and Uncertainty in the Post Keynesian Approach

The theoretical developments briefly outlined in the previous section draw upon Ellsberg's examples to propose an alternative axiomatic structure to the subjective expected utility model. This alternative is characterized by the rejection of the assumption of a single subjective prior probability, thereby extending the explanatory content of the theory to encompass situations of deliberate violation of a fundamental axiom of the prevailing standard theory.

The original theory is thus generalized following a process that is essentially the same as that identified by Tjalling Koopmans when he, in turn, supported Savage's developments because they allowed for a generalization of von Neumann and Morgenstern's expected utility theory. Although sometimes characterized as 'non-Bayesian,' the new approach can actually be considered as a generalization of the Bayesian approach in that, while it challenges one of its pillars, it accepts the general framework provided by the other pillars. With notable exceptions such as Gilboa and Schmeidler (1995) and Binmore (2009), the assumption that the state space is known to the decision-maker in every context is not challenged.[6]

[5] Gilboa and Marinacci (2013) and Machina and Siniscalchi (2014) summarize the significance of this literature, usually identified as the 'ambiguity aversion' literature.

[6] Bradley (2017) provides a methodologically grounded analysis of the directions that a decision theory aiming for greater realism should take, without abandoning the prescriptive perspective typical of the Bayesian mainstream.

In the tradition of thought that emphasizes 'Keynesian' uncertainty, the passage just outlined is regarded as inadequate. It may be useful for dealing with ambiguous decision contexts, but not for those that Keynes had in mind when he developed his theory as an alternative to that of the 'classical' economists. In particular, the American economist Paul Davidson (1930–2023) pointed out that the critical factors that Keynes had in mind when analyzing the (mal)functioning of economic systems in the 1930s—specifically, liquidity preference and animal spirits—are inherent in his notion of uncertainty. As discussed in Chap. 3, Keynes noted that there are instances where the attempt of monetary policy to induce interest rate reductions aimed at stimulating private investment is ineffective. In circumstances characterized by uncertainty, such as when stock values are perceived to be linked to the conventional behavior of financial agents rather than to fundamental values, individuals who hold money for precautionary or speculative purposes do so more out of dislike of the uncertain environment than out of expectations about changes in the interest rate. Similarly, entrepreneurs have no incentive to undertake new investments, even if they are convinced of the profitability of such endeavors.

In both cases, the idea is that the agents described by Keynes are decision-makers who cannot be represented in terms of modern rational choice theory. This is due to the fact that they "cannot conceive of a complete list of consequences that will occur in the future" and therefore their degrees of belief cannot be identified with probabilities, because "the available evidence is insufficient to establish a probability" (Davidson, 1991, p. 134). The fact that there is not enough evidence to interpret degrees of belief as probabilities precludes the usual process of ordering actions and selecting the option that maximizes the decision-maker's utility. For this reason, Keynes attached considerable importance to the 'state of long-run expectation' when examining the factors relevant to decisions to invest in new capital. He believed that in the early 1930s expectations had created conditions in which financial markets did not justify investment except for speculative purposes. Moreover, the increase in the liquidity premium had hindered the ability of monetary policy to induce a reduction in the interest rate and, consequently, to generate growth in the economic system.

From a decision-theoretic perspective, Davidson's critique mainly concerns the first pillar of Bayesian theory. This critique resonates with that formulated by Shackle in the years following von Neumann and Morgenstern's systematization of decision theory under risk. As noted in Chap. 6, Shackle had endeavored to develop a non-probabilistic approach to decision theory under uncertainty, an attempt that received considerable attention from Arrow (1951), before being dismissed as lacking a convincing axiomatic foundation. Following Shackle, post Keynesian scholars emphasize that the really important decisions are those in which an individual cannot completely ignore "the fear of tragedy because of unforeseeable changes during the time between choice and outcome" (Davidson, 1991, p. 136). For many routine decisions, or in what Savage termed a small world, the assumption that the list of states of nature to be considered is exhaustive may be a useful simplification, because it aligns well with the idea that the decision environment is not particularly different from the past. However, as Davidson has argued, decisions about liquidity and investment are typically influenced by contingencies that are difficult to predict in advance.

Therefore, novelty and surprise must be part of decision-makers' imagination as they envision the future, and crucial decisions are more accurately represented as being made in a large world, a context for which Savage himself acknowledged that a narrow probabilistic scheme of reasoning is ill-suited.[7]

To this analysis, the Post Keynesians have added a critical view of the significance of Ellsberg's critique and the resulting theories that propose alternative decision criteria, as outlined in the previous section. Post Keynesian scholars acknowledge the relevance of Ellsberg's contribution in accurately pointing out an insufficiency of the dominant approach, but they contend that the context of ambiguity remains inadequate for the study of Keynesian uncertainty. If one were to restrict oneself to Ellsberg's considerations, uncertainty would be confined to issues of incomplete information and unreliable probability functions; but according to the Post Keynesians, the real Keynesian point is that certain information is simply not available at the time of the decision (Dequech, 2000).

According to this interpretation, when Keynes rejected the idea that there are "computable" probabilities necessary for the calculation of expected utility, he did so because he believed that one could not fully specify the list of contingencies that might affect the outcome of the decision. From a post Keynesian perspective, modern decision theory, while critical of Savage, has continued to reason as if the world were an urn, albeit with unknown probabilities. But what attracts the attention of real-world decision-makers in making crucial economic decisions are potentially unforeseen events. The irreducibility of Keynesian uncertainty to risk is thus based not so much on a rejection of the sure-thing axiom/principle—Ellsberg's focus—as on the idea that the decision-making context must take account of unforeseen contingencies.[8]

To summarize, scholars such as Davidson contend that the standpoint of those decision theorists who have been critical of the dominant approach remains inadequate. By accepting the first pillar of the Bayesian approach, modern uncertainty theories—whether employing non-additive probabilities or multiple probabilities—continue to place uncertainty within a strictly probabilistic framework. Conversely, true Keynesian uncertainty would not be measurable in terms of probabilities and would necessitate an entirely different approach. This long-standing argument about uncertainty—sometimes termed 'fundamental Keynesianism'—has recently gained support even among decision theorists. A notable proponent of this approach is the English decision theorist Ken Binmore, whose arguments have departed from the mainstream to align with the aforementioned perspective. In a paper aptly

[7] That this entails an ontological commitment, the view that the real world is characterized as an environment that cannot be perfectly known in advance because of the way it is structured—an open-ended view of the world—is a relevant part of the argument that we do not address here. Indeed, we have focused mainly on the epistemic side of our subject matter, leaving many relevant ontological aspects unexplored. Notably, Lawson (2006) regards fundamental uncertainty as a key distinguishing element of the ontology of heterodox thinking. Davis (2024) offers an example of how this ontological commitment can foster an alternative view of the individual economic agent.

[8] A defense of the importance of Ellsberg's analysis for a Keynesian viewpoint is provided in Basili and Zappia (2010).

titled "Rational decisions in large worlds," Binmore (2007, p. 25) has advanced the following argument:

> Behavioral economists speak of ambiguity rather than uncertainty, but I see this as a retrograde step, since it suggests that the difficulty with uncertain situations is only that the decision-maker may be in two minds about what probability to assign to an uncertain event …. But are we so sure that it always makes sense to assign subjective probabilities to all uncertain events?

However, after drawing attention to the limitations of Bayesianism and focusing on those situations in which decision-makers are unable to decide mathematically "undecidable propositions," Binmore (2007, p. 40) proposes a modeling framework for large worlds. He argues that, as a theorist, he is unwilling "to abandon any hope of modeling rational choice in worlds that are as complicated as we are ourselves." In contrast, the radical alternative proposed by post Keynesian scholars faces the risk of being overlooked because it lacks a formal framework capable of detailing alternative norms of behavior for decision-makers under conditions of true/fundamental uncertainty. Indeed, we have mentioned Shackle's name again because his non-probabilistic approach is still the only one that attempts to give a formalized representation to the considerations emphasized by the Post Keynesians.[9]

Moreover, the interpretation just outlined seems to classify Keynesian uncertainty as a form of complete ignorance, or at least similar in substance to Knight's uncertainty. Even among Keynesian scholars, this interpretation is not unanimously shared. As discussed in Chap. 3, Keynes's probabilistic analysis was very sophisticated. When employed as a key to interpreting the uncertainty underlying the *General Theory*, it becomes possible to identify a gradation of uncertain situations, rather than a sharp dichotomy between situations of risk and situations of uncertainty. Keynes's attitude toward the feasibility of probabilistic calculation appears to be more receptive than that attributed to him by the Post Keynesians. Specifically, the assertion that there may not be 'calculable' probabilities, as Keynes noted in the passage we have quoted at length from his 1937 *Quarterly Journal of Economics* article, does not entail an absolute rejection of probabilistic reasoning. The non-numerical probabilities introduced by Keynes in the *Treatise on Probability* are not intended to characterize situations in which probabilities are simply nonexistent; rather, they are probabilities of a special kind that can be used even in the most uncertain situations. These probabilities can be conceptualized as formally analogous to interval-valued probabilities or probabilities that do not respect the standard properties, such as additivity.

Furthermore, as we noted in the preceding chapter when illustrating Ellsberg's approach, Keynes's probabilistic framework nevertheless serves as the basis for those probabilistic developments that seek to capture the individuals' attitude toward uncertainty in a more comprehensive manner, as opposed to those that reject the value of

---

[9] As noted in Hamouda and Rowley (1996) and elaborated in Basili and Zappia (2009), the non-probabilistic approach that Shackle proposed in the late 1940s and then defended in the following decades—although increasingly isolated and eventually forgotten by decision theorists—has found in some approaches proposed by scholars of fuzzy measures and imprecise probabilities a possible revival. For a recent reappraisal of Shackle's model, see Katzner (2023).

such developments altogether. And Keynes would likely have accepted a consideration that Ellsberg ([1962] 2001, p. 21) reported from one of the subjectivist scholars we have mentioned as a proponent of an axiomatic approach to imprecise probabilities, the English statistician Irving Good (1952, p. 107), who stated: "There may be situations in which it is best to behave irrationally, but if there are, it should be decided rationally." This line of reasoning also seems to align with Binmore's emphasis on trying to make sense of decisions in the context of a large world without abandoning formalized modeling.

7.4 The Great Recession and Keynesian Uncertainty

We have seen in the previous sections that there are critical approaches to the theory of choice under conditions of uncertainty that dominated the second half of the twentieth century, and that these approaches refer more or less explicitly to a Keynesian view of uncertainty. But a substantial rehabilitation of Keynesian themes has also emerged in recent years for contingent historical reasons. In the early 2000s, the global economic system experienced a prolonged period of crisis, often compared to the Great Depression of the 1930s, and therefore referred to as the Great Recession. A series of events that initially appeared to be confined to the housing market in the United States, threw the entire financial system went into crisis, leading first to a collapse in stock market valuations, then to a contraction in the credit market, and finally to a long period of recession. The gross national product of the United States and, in a cascade, of a number of other Western countries experienced growth rates that were initially negative and then very low for many quarters. In mid-2007, global growth in the Organization for Economic Co-operation and Development (OECD) countries was expected to be above 2%, in line with the previous decade, but growth actually remained negative for almost two years and then largely stagnated.

What had appeared to be a long period of growth in the US housing market turned out to be a genuine speculative bubble, based on sophisticated mechanisms of securitization of financial instruments issued against the issuance of mortgages. Irrespective of the actual mechanism, which involved the pooling of individual mortgages and the subsequent issuance of securities linked to those mortgages, securities backed by high credit ratings were offered in the market to traders who in turn resold them to other traders, essentially preventing the real underlying—increasingly consisting of mortgages offered to potentially insolvent consumers—from being linked to the security representing it. The rational justification was that the new instruments would have allowed a better assessment of risks and, at the same time, a more effective hedge against possible defaults linked to the underlying real assets, the mortgages. In reality, as the financial actors themselves were later forced to admit, they were confronted with exactly the opposite phenomenon, namely an increasing inability to correctly assess risks, a phenomenon that was eventually termed risk mispricing.

The financial collapse and the ensuing Great Recession revealed that the supposed ability of competitive markets to regulate themselves was illusory. After the bankruptcy of a first major player—the financial services firm Lehman Brothers—monetary authorities first intervened to save lenders from bankruptcy, including through nationalizations, and then pursued expansionary policies for years on a scale not seen since World War II. Yet, many scholars continued to argue that the problem was related to informational issues and inappropriate incentives: mispricing, it was argued, should be countered by better risk management, which in any case should be entrusted to financial agents on the basis of more accurate risk models and improved incentive design. According to this approach, even after such a globally catastrophic event, the observed financial data could still be interpreted as temporary deviations from the presumed ability of market participants to correctly assess the fundamental values of investments and related financial assets.

Robert Skidelsky, author of the most authoritative biography of Keynes, recalls that at the beginning of this century, when he was publishing the third volume of his monumental biography, interest in Keynes had largely waned. Most American economists were opposed to fiscal stimulus on the basis of an argument that Keynes had fought against in the 1930s, namely, that the use of tax revenues to finance public investment would at best have short-term effects, but would then prove to provoke the 'crowding-out' of private investment. According to this largely agreed view, it was necessary not only to reduce taxation, but also to limit the weight of the state in the economy in general, including in the form of further deregulation of private activity. In this regard, Skidelsky (2009, pp. 48–49) quotes the main proponent of the theory of efficient financial markets, the (soon-to-be) Nobel laureate Eugene Fama of the University of Chicago, who, despite the crisis, commented on the attempt by the Federal Reserve to provide ample liquidity to the system: "Bailouts and stimulus plans are funded by issuing more government debt … The added debt absorbs savings that would otherwise go to private investment. In the end, despite the existence of idle resources, bailouts and stimulus plans … just move resources from one use to another." Skidelsky also reports the harsh reaction to this approach from one of the few defenders of Keynesian policies, the Nobel laureate Paul Krugman, who commented on his blog:

> If there was one essential element in the work of John Maynard Keynes, it was that [...] the fact that spending equals output, or equivalently that saving equals investment, does *not* imply that there is always enough spending to fully employ the economies, that there is always enough investment to make use of the savings the economy would have had if it were at full employment.

But Krugman's considerations were considered a minority viewpoint when the crisis began.[10]

Indeed, according to the theory of efficient financial markets, the market prices of financial assets represent the best estimate of the risks associated with holding those assets. Not only is it assumed that the differences between the valuations of different assets reflect the different risks, but also that the risks are correctly reflected in

[10] Skidelsky (2009, p. 49) also cites Krugman's discomforting conclusion: "it is deeply depressing to find, not that people like Eugene Fama disagree with Keynes's conclusions … but that they are obviously completely unaware of the whole argument."

the prices. "The risk characteristic of financial markets," according to a paper from those years by the British Financial Services Authority cited by Skidelsky (2009, p. 38), "can be inferred from mathematical analysis, delivering robust quantitative measures of trading risk." In this context, "harmful risk taking" is constrained by the discipline imposed by markets, and financial innovations can be assumed to be "beneficial since market competition would winnow out innovations which did not deliver value added."

Underlying all arguments about the efficiency of financial markets and their ability to allocate savings correctly—an allocation that government intervention would only hinder—is a reliance on risk assessment models that exclude any consideration of significant uncertainties. These models assume that risks are always assessable in a probabilistic sense and that unexpected events are confined to the 'tails' of probability distributions around the mean, so that these events are assigned such low probabilities that they end up being irrelevant to decision. But the real issue—whether the mean of the probability distribution is reliable or not—is not addressed. To simplify, Skidelsky reminds us, it is as if financial risks were valued in the same way as risks associated with life insurance, considered as the typical sector in which the amount of data is so overwhelming to let firms comfortably operate. We know from Chap. 3 that Knight had used precisely the theme of insurability to explain in practical terms the necessity to distinguish risk from uncertainty. And we saw in Chap. 5 that even Bruno de Finetti—the most radical of the subjectivists—while denying in principle that Knight's reasoning could provide a sound basis for decision-making under uncertainty, had admitted that there are situations in which the divergent assessments of insurance operators can lead to incompleteness in insurance markets.

But it is the reference to Keynes's reasoning that clearly illustrates the difficulty of considering financial markets as efficient allocation mechanisms. It is not just that the probabilities associated with events potentially disruptive to the stability of the system are so low as to be irrelevant, but that the way in which these probabilities are assigned fails to capture the specificities of how markets operate. Depending on the degree of confidence in the system's ability to function properly, the probabilities that are assigned to events are more or less accurate. And when confidence in the ability to determine accurate probabilities is low, the usual decision-making apparatus does not work.

This is not to say that the probabilities of unexpected events do not exist, as advocated in the extreme version of post Keynesianism in the style of Davidson. Rather, it is a matter of recognizing that standard probabilistic reasoning is inadequate, because the more uncertain the situation is perceived to be, the more likely it is that only qualitative probability comparisons can be made. And this can be represented by the idea that the probabilities that can be used are non-numerical, as opposed to the precise probabilities that a strictly Bayesian approach always assumes to be available.

On this point, it is worth noting that Ellsberg ([1962] 2001, p. 165) characterized himself and those working in the vein of Keynes's probability theory as "less exacting Bayesians," to indicate that a notable tradition of thought remained anchored in the idea that some form of probabilistic calculation was necessary. But Ellsberg also acknowledged that Shackle's extreme position could not be ignored, at least when ambiguity is so severe to generate extremely low confidence in the available probabilistic estimates. Indeed, Ellsberg also quoted Shackle very favorably when he argued that he conceived of expectations "as an act of creative imagination and not of 'rational' calculation; for calculation is impossible when data are incomplete, and in the face of ignorance, rationality is a mere pretense" (Shackle, 1955, p. 74).

It is not possible here to explore further the question of whether Keynes belonged more to the first or second characterization offered by Ellsberg. Nor is it possible to explore whether the caution against accepting overly ambitious answers should be interpreted as an outright rejection of the modest guidance offered by probability theory, as suggested by the Post Keynesians. But certainly Keynes was striving for a theory of 'reasonable' behavior under conditions of uncertainty, showing a definite interest in discussing the issue of what decision criteria can be devised when those suggested by the application of a strict notion of rationality are untenable. His rejection of the reduction of uncertainty to risk could therefore have been a guide to a more nuanced understanding of how financial markets actually work, something that was instead ignored in the years leading up to the crisis.

7.5 Concluding Remarks

At the time of the publication of his book, with the crisis still unfolding, Skidelsky (2009, pp. 50–51) concluded his analysis by arguing that it was not difficult to imagine what Keynes would have thought about the performance of the economic system in those years. Keynes might not have predicted that the financial collapse would come when it did, but he would certainly have thought that all the conditions were in place for the collapse to occur. It would have been evident to Keynes that the economic authorities had failed to take substantial measures to control the booming economy. Furthermore, he would have discerned that a system comprising financial operators who believed that they could 'price in' any kind of risk was doomed to failure.

Much of the economics profession had encouraged the development of models in which probabilistic assumptions about future events were considered reliable in all contexts, thereby devaluing the importance of keeping uncertainty separate from risk. On this basis, there had been a shift from Markowitz-like theories of efficient portfolios to general equilibrium models of financial markets. These models ascribed to financial markets the kind of efficiency properties that theoretical economists of the 1950s, such as Arrow and Debreu, had imagined only for stylized models of the economy based on extremely restrictive assumptions, such as the existence of markets for exchanges contingent on any kind of event. Not only had the considerations of Chap. 12 of the *General Theory* on the instability of financial markets been completely forgotten, but in the new sophisticated models there was not even a trace of Savage's caution about the need to differentiate applications of the theory in small and large worlds.

In a manner analogous to the Great Depression of the 1930s, which necessitated the kind of analysis that Keynes provided in the *General Theory*, the recent Great

Recession justified a return to Keynes. In both cases, the guidance for economic policy comes from an analysis of economic systems that places emphasis on the role of a specific form of uncertainty, Keynesian uncertainty. It is this idea of uncertainty that has constituted the central theme of this volume.

References

Arrow, K. J. (1951). Alternative approaches to the theory of choice in risk-taking situations. *Econometrica, 19*, 404–437.

Basili, M., & Zappia, C. (2009). Shackle and modern decision theory. *Metroeconomica, 60*, 245–282.

Basili, M., & Zappia, C. (2010). Ambiguity and uncertainty in Ellsberg and Shackle. *Cambridge Journal of Economics, 34*, 449–474.

Binmore, K. (2007). Rational decisions in large worlds. *Annales D'économie Et De Statistique, 86*, 25–41.

Binmore, K. (2009). *Rational decisions*. Princeton University Press.

Bradley, R. (2017). *Decision theory with a human face*. Cambridge University Press.

Davidson, P. (1991). Is probability theory relevant for uncertainty? A post Keynesian perspective. *Journal of Economic Perspectives, 5*, 129–143.

Davis, J. B. (2024). *Identity, capabilities, and changing economics*. Cambridge University Press.

Dequech, D. (2000). Fundamental uncertainty and ambiguity. *Eastern Economic Journal, 26*, 41–60.

Dow, J., & da Costa Werlang, S. R. (1992). Uncertainty aversion, risk aversion, and the optimal choice of portfolio. *Econometrica, 60*, 197–204.

Ellsberg, D. ([1962] 2001). *Risk, ambiguity and decision*. Garland Publishing.

Gigerenzer, G. (2008). *Rationality for mortals: How people cope with uncertainty*. Oxford University Press.

Gilboa, I. (2009). *Theory of decision under uncertainty*. Cambridge University Press.

Gilboa, I. (2015). Rationality and the Bayesian paradigm. *Journal of Economic Methodology, 22*, 312–334.

Gilboa, I. (2025). Decision under uncertainty: State of the science. *Annual Reviews in Economics, 17*, forthcoming.

Gilboa, I., & Marinacci, M. (2013). Ambiguity and the Bayesian paradigm. In D. Acemoglu, M. Arellano, & E. Dekel (Eds.), *Advances in economics and econometrics: Vol. 1*. Cambridge University Press.

Gilboa, I., Postlewaite, A. W., & Schmeidler, D. (2008). Probability and uncertainty in economic modeling. *Journal of Economic Perspectives, 22*, 173–188.

Gilboa, I., & Schmeidler, D. (1989). Maxmin expected utility with non-unique prior. *Journal of Mathematical Economics, 18*, 141–153.

Gilboa, I., & Schmeidler, D. (1995). Case-based decision theory. *Quarterly Journal of Economics, 110*, 605–639.

Good, I. J. (1952). Rational decisions. *Journal of the Royal Statistical Society, Series B, 14*, 107–114.

Hamouda, O. F., & Rowley. R. (1996). *Probability in economics*. Routledge.

Hansen, L. P., & Sargent, T. J. (2001). Acknowledging misspecification in macroeconomic theory. *Review of Economic Dynamics, 4*, 519–535.

Hirshleifer, J., & Riley, J. C. (1992). *The analytics of uncertainty and information*. Cambridge University Press.

Kahneman, D., & Tversky, A. (1979). Prospect theory: An analysis of decision under risk. *Econometrica, 47*, 263–291.

Katzner, D. W. (2023). The problem with probability. *Journal of Post Keynesian Economics, 46*, 379–399.

Keynes, J. M. (1937). The general theory of employment. *Quarterly Journal of Economics, 51*, 209–223.
Lawson, T. (2006). The nature of heterodox economics. *Cambridge Journal of Economics, 30*, 483–505.
Machina, M. (1987). Choice under uncertainty: Problems solved and unsolved. *Journal of Economic Perspectives, 1*, 121–154.
Machina, M., & Siniscalchi, M. (2014). Ambiguity and ambiguity aversion. In M. Machina & W. K. Viscusi (Eds.), *The handbook of the economics of risk and uncertainty*. Elsevier.
Mirowski, P., & Nik-Khah, E. (2017). *The knowledge we lost in information: The history of information in modern economics*. Oxford University Press.
Petracca, E. (2021). On the origins and consequences of Simon's modular approach to bounded rationality in economics. *European Journal of the History of Economic Thought, 28*, 708–732.
Rizzello, S. (2002). Herbert Simon's heritage. *International Review of Economics, 2*, 289–304.
Schmeidler, D. (1989). Subjective probability and expected utility without additivity. *Econometrica, 57*, 571–587.
Shackle, G. L. S. (1955). *Uncertainty in economics*. Cambridge University Press.
Simon, H. A. (1959). Theories of decision-making in economics and behavioral science. *American Economic Review, 49*, 253–283.
Skidelsky, R. (2009). *The return of the master*. Public Affairs.
Wakker, P. (2010). *Prospect theory: For risk and ambiguity*. Cambridge University Press.
Zappia, C. (2018). Rationality under uncertainty: Classic and current criticisms of the Bayesian viewpoint. *European Journal of the History of Economic Thought, 25*, 1387–1419.

The manufacturer's authorised representative in the EU is Springer Nature Customer Service Centre GmbH, Europaplatz 3, 69115 Heidelberg, Germany. If you have any concerns regarding our products, please contact ProductSafety@springernature.com

Printed and bound by CPI Group (UK) Ltd, Croydon, CR0 4YY
01/10/2026
02225978-0019